SYNTHETIC GENE NETWORK

Modeling, Analysis, and Robust Design Methods

SYNTHETIC GENE NETWORK
Modeling, Analysis, and Robust Design Methods

Bor-Sen Chen

Laboratory of Control and Systems Biology
Department of Electrical Engineering
National Tsing Hua University
Hsinchu, Taiwan

and

Yu-Chao Wang

Institute of Biomedical Informatics
National Yang-Ming University
Taipei, Taiwan

CRC Press
Taylor & Francis Group
Boca Raton   London   New York

CRC Press is an imprint of the
Taylor & Francis Group, an **informa** business

A SCIENCE PUBLISHERS BOOK

CRC Press
Taylor & Francis Group
6000 Broken Sound Parkway NW, Suite 300
Boca Raton, FL 33487-2742

International Standard Book Number: 978-1-4665-9269-8 (Hardback)

**Visit the Taylor & Francis Web site at
http://www.taylorandfrancis.com**

**CRC Press Web site at
http://www.crcpress.com**

**Science Publishers Web site at
http://www.scipub.net**

Preface

Synthetic biology is an engineering discipline that builds on our mechanistic understanding of molecular biology to program microbes or mammalian cells to carry new functions. In the past decade of development, significant progress has been made in designing biological parts and assembling them into genetic circuits to achieve basic functionalities. Synthetic biology currently focuses on the rational construction of biological systems based on engineering principles. Such predictable manipulation of a cell for these biological systems requires modeling, systems analysis and design as well as experimental techniques to work together. The modeling component of synthetic biology allows one to design genetic network and analyze its expected behavior, and the experimental component merges model with real systems by providing quantitative data and sets of available biological 'parts' to construct genetic networks. Since there are several intrinsic kinetic fluctuations such as splicing and mutation as well as extrinsic disturbances such as thermal noise and upstream interference, stochastic dynamic models are introduced for genetic networks *in vivo*.

Based on stochastic dynamic model, kinetic activities are identified for biological parts to redefine the libraries by least square parameter estimation method via experimental data. Then according to the gene network topology and design specifications, several robust design algorithms are systematically developed to select adequate biological parts from their corresponding libraries for the synthetic gene network to satisfy the prescribed design specifications under intrinsic fluctuations and extrinsic disturbances *in vivo*. When the volume of libraries becomes very large, the genetic algorithm (GA)-based and evolutionary algorithm (EA)-based methods are also developed to efficiently search for adequate biological parts from the corresponding libraries to engineer a desired synthetic gene network.

In order to achieve a desired behavior or function under intrinsic and extrinsic molecular noises, the H_2 optimal tracking and H_∞ robust disturbance attenuation have been provided as two design specifications for synthetic gene networks in this book. In order to solve robust synthetic gene network design problem in H_2 optimal tracking, H_∞ robust filtering or mixed H_2/H_∞ robust tracking design, we need to solve a corresponding Hamilton-Jacobi inequality (HJI)-constrained optimization problem. Therefore, the global linearization technique or fuzzy interpolation method is employed to interpolate several local linear systems to approximate the nonlinear stochastic synthetic gene networks. In this situation, a set of linear matrix inequalities (LMIs)-constrained optimization problem can replace a HJI-constrained optimization problem so that it can be easily solved by using the LMI toolbox in Matlab in the design procedure of robust synthetic gene network.

In the IC industry, due to high complexity and difficulty in very large scale integrated (VLSI) circuits, which can not be imaged about 2 or 3 decades ago, system design companies (like Intel) and system implementation companies (like Taiwan Semiconductor Manufacturing Company, TSMC) perform the products of VLSI circuits by division of labor. In the future, systems designers should cooperate with implementation companies to produce complex synthetic gene networks. If this is the case, the development of systematic design tools is an important topic for synthetic gene networks. Therefore, the proposed robust optimal tracking design methods have potential applications to synthetic gene network design in the near future.

We would like to thank Chih-Hung Wu, Chih-Yuan Hsu, and all the members in our lab for their contributions to this book.

Bor-Sen Chen
Yu-Chao Wang

Contents

1

General Introduction

The main goal of the nascent field of synthetic biology is to engineer an artificial gene circuit and then insert it into the host cell to perform new tasks. One useful analogy to conceptualize both the goal and methods of synthetic biology is the computer engineering hierarchy. At the bottom of the hierarchy of synthetic biology are DNA, RNA, proteins, and metabolites (including lipids and carbohydrates, amino acids, and nucleotides), analogous to the physical layer of transistors, capacitors, and resistors in computer engineering (Andrianantoandro et al. 2006). The next layer, the device layer, comprises biochemical reactions that regulate the flow of information and manipulate physical processes, equivalent to engineered logic gates that perform computations in a computer. At the module layer, synthetic biologists use a diverse library of biological devices to assemble complex pathways that function like integrated circuits (ICs).

However, building biological systems entails a unique set of design problems and solutions. Biological devices and modules are not independent objects, and they are not built in the absence of a biological context. Biological devices and modules of synthetic biology typically function within cellular environments. When synthetic biologists engineer the devices or modules, they do so using the resources and machinery of the host cell, but in the process also modify the cells themselves. A major concern in this process is our present inability to fully predict the functions of even simple devices in the engineered cells and construct systems that perform complex tasks with precision and reliability (Andrianantoandro et al. 2006). The lack of predictive power stems from several sources of uncertainty, some of which signify the incompleteness of available information regarding inherent cellular characteristics. The effects of gene expression noises, uncertain initial conditions, mutations, cell death, undefined and changing extracellular environments, and interactions with cellular contexts currently hinder us from engineering biological systems with the confidence that we can engineer computers to do specific tasks.

The emerging discipline of synthetic biology is regarded as the rational engineering of biological processes for practical use. It is a discipline at the intersection of protein and genetic engineering with systems biology and has the ambitious goal of extending current biotechnology to large-scale gene circuits. Hence, synthetic biology can be regarded as a vehicle for understanding relationships between biology and engineering. In the last decade, synthetic biology has shown potential for generating practically useful systems while serving as an approach to implement biological circuits as "design principles" (Chin 2006, Haseltine and Arnold 2007). Some genetic circuits are constructed according to these design principles, such as toggle switches (Gardner et al. 2000, Atkinson et al. 2003, Kramer et al. 2004, Ham et al. 2008), transcriptional cascades (Hooshangi et al. 2005), pulse generators (Basu et al. 2004), genetic counters (Friedland et al. 2009), oscillators (Atkinson et al. 2003, Goh et al. 2008, Stricker et al. 2008, Tigges et al. 2009), logic evaluators (Rinaudo et al. 2007, Win and Smolke 2008), sensors (Kobayashi et al. 2004, Win and Smolke 2007) and cell-cell communicators (Kobayashi et al. 2004, Basu et al. 2005, Pai et al. 2009). Synthetic biologists provide a concept that programmable cells can be constructed by designing appropriate interfaces that couple engineered gene circuits to the regulatory circuitry of the host cell (Kobayashi et al. 2004). The behavior of a complicated system is predicted from the characteristics of its elementary parts (Hasty et al. 2002, Sprinzak and Elowitz 2005, Andrianantoandro et al. 2006, Simpson 2006). From this viewpoint, well-characterized biological parts and an accurate mathematical model are beneficial for the construction of a more complex biological system (Guido et al. 2006).

For synthetic biologists, it is appealing to engineer a gene circuit with some desired behaviors for practical applications. For example, the field has yielded several technological applications and provided new avenues for drug manufacture (Withers and Keasling 2007, Weber et al. 2008), biofabrication (Basu et al. 2005), therapeutics (Lu and Collins 2007, Lu and Collins 2009, Xie et al. 2011), and biofuel production (Lee et al. 2008, Steen et al. 2008, Waks and Silver 2009). However, synthetic circuit construction is still at a time-consuming post-hoc tweaking phase because of having to work with a limited set of unsuitable parts. Unlike other engineering disciplines, synthetic biology has not developed a scalable and reliable method to find solutions. Instead, emerging applications are most often kludge that can be used only in individual special cases (Arkin and Fletcher 2006). Hence, traditional methods for gene circuit construction typically build a prototype of the circuit, and then carry out a laborious process of trial-and-error to modify components before the desired performance is achieved, or use directed evolution (Yokobayashi et al. 2002, Haseltine and Arnold 2007, Hawkins et al. 2007, Dougherty and Arnold 2009) to mutate and then screen out the functional designs. The drawbacks of these methods include lack of predictability, and the long time and great effort required to obtain a functioning circuit. We seek to extract interchangeable biological parts from living systems that can be tested, validated as construction units, and reassembled to create devices that can have analogs in living systems (Gibbs 2004). Therefore, one engineering goal might be to create standard biological parts. Fortunately, a Registry of Standard Biological Parts initiated at MIT maintains and distributes thousands of BioBrick standard biological parts (http://www.partsregistry.org). Synthetic biologists can rapidly construct a novel genetic circuit by assembling these BioBrick parts that provide the physical composition (Canton et al. 2008). Although constant efforts are being made by the community to improve the reliable parts, engineering a synthetic gene circuit to track a desired behavior still remains an acute problem because most of the available biological parts are inadequately characterized (Ellis et al. 2009). In order to employ quickly these BioBrick parts in engineering a synthetic gene circuit, synthetic biologists need more suitable biological libraries for engineering a gene circuit.

The design goal of synthetic biology is to assemble biological parts to create a functional gene circuit. Even though recent construction of synthetic gene networks has demonstrated the feasibility of synthetic biology, as described above, the design of gene networks remains difficult and the most newly designed gene networks cannot function properly. These design failures are mainly due to both intrinsic perturbations (e.g., gene expression noises, splicing, mutation, and evolution) and extrinsic disturbances (e.g., as interference from upstream molecules, and changing extra-cellular environments) (Andrianantoandro et al. 2006, Batt et al. 2007). Therefore, optimal design of a robust synthetic gene network that can tolerate intrinsic parameter perturbations, attenuate extrinsic disturbances, and function properly in a host cell is an important topic in synthetic biology. In this case, mathematical modeling and systems theory can be useful tools that help synthetic biologists predict and implement gene circuits. Recently, several robust methods for the design of synthetic gene networks have been implemented in the construction of synthetic gene networks to achieve desired behaviors under design specifications (Chen et al. 2009, Chen and Wu 2009, Chen and Chen 2010, Chen and Wu 2010, Chen et al. 2011a, Chen et al. 2011b, Chen and Lin 2013). More recently, robust design methods for synthetic gene networks have been proposed by developing a systematic method to efficiently employ promoter libraries to improve the engineering of synthetic gene networks with desired behaviors (Wu et al. 2011a, Wu et al. 2011b). These modeling and robust design methods will be introduced in this book.

In this book, the mathematical models and design specifications are introduced in Chapter 2. Actually, many molecular-level processes of synthetic gene networks are deeply rooted in statistical mechanical behaviors of so-called nanoscale biochemical systems. A nonlinear stochastic model is presented to analyze the dynamic properties of synthetic gene networks with parameter uncertainties, external disturbances and functional variations in the host cell. Then several design specifications, including H_2 optimal tracking and H_∞ disturbance filtering, are introduced to guarantee that synthetic gene network could work with desired state behaviors under intrinsic parameter fluctuations, external disturbances and functional variations in the host cell. In Chapter 3, after the mathematical models and design specifications have been introduced, several robust synthetic design methods are proposed for robust synthetic gene networks to meet these design specifications and function properly in spite of intrinsic parameter fluctuations and extrinsic disturbances. To avoid directly solving nonlinear stochastic stabilization and the disturbance filtering design problems of robust synthetic gene network, the global linearization and fuzzy interpolation techniques are employed to transform a nonlinear stochastic gene network to a set of local linearized gene networks to simplify the design procedure so that the robust synthetic gene network design problem could be solved efficiently.

In light of natural selection on traits best-suited for environmental change being an important mechanism in evolution (Ayala 2007), the question arises whether a similar strategy can be adapted for gene network design. In Chapter 4, inspired by biological evolution, two network evolutionary methods based on genetic algorithm (GA) and evolutionary algorithm (EA) are proposed for robust synthetic gene network design. In order to mimic the naturally occurring biological systems in evolutionary process, the fitness function is selected to be inverse proportional to the tracking error so that evolutionary kinetic parameters of synthetic gene network can achieve the optimal desired tracking behavior via the maximization of fitness with all speed to mimic the evolution process of a gene network. If the adaptations of kinetic parameters of synthetic gene network are reflected by the proposed network evolutionary algorithm to achieve the optimal fitness, the evolutionary gene network will track the desired biological function in spite of intrinsic parameter fluctuations and extrinsic noise and will behave more robustly inside a living cell.

Some current promoter libraries have been developed for synthetic gene networks. However, it is still difficult to directly select promoters from these promoter libraries for engineering a synthetic gene network to achieve desired behaviors because these promoters still lack promoter activities which can be easily selected for synthetic biologists. Therefore, there exists no efficient method to engineer a synthetic gene network with some desired behaviors by selecting adequate promoters from these promoter libraries. In Chapter 5, for the convenience of design, we will redefine the component activities for promoter components or promoter-RBS components from their regulatory expression data in these conventional component libraries so that they can be easily selected for the robust synthetic gene networks. After the promoter and promoter-RBS libraries have been redefined based on component activities of dynamic regulatory equations, in Chapters 6 and 7, we also develop a systematic method to efficiently select adequate promoters/promoter-RBSs from these redefined promoter/promoter-RBS libraries so that the synthetic gene network with desired behaviors can be implemented easily. Based on H_2 optimal tracking and H_∞ disturbance filtering performance, an adequate promoter set can be efficiently selected from the redefined promoter libraries for a synthetic gene network to achieve multi-objective H_2/H_∞ reference tracking design. In addition, two synthetic gene network design examples based on promoter-RBS libraries, i.e., biological filter and transistor designs, are given in Chapter 7.

Recently, a simple synthetic device was engineered in a cell, and several cells were then combined, so that their connections allowed the construction of a more complex synthetic gene network, i.e., so-called multi-cellular engineered gene networks. This approach not only uses cellular consortia as an efficient way of engineering complex gene networks, but also demonstrates the great potential for reutilization of small parts of the gene network. Therefore, collective rhythms of gene regulatory networks have been an important subject of considerable interests for biologists and theoreticians, in particular the synchronization of dynamic cells mediated by intercellular communication. In this situation, synchronization of a population of synthetic gene networks is an important design in practical applications in future, because such a population of synthetic gene networks distributed over different host cells need to exploit molecular phenomena simultaneously in order to emerge a biological phenomenon. However, this intercellular communication and synchronization may be corrupted by intrinsic kinetic parameter fluctuations and extrinsic environmental noise. In Chapter 8, the synchronization robustness criterion is defined for the stochastic coupled synthetic gene networks in spite of intrinsic parameter fluctuation and extrinsic disturbance. If the synchronization robustness criterion is violated, some control schemes are designed to improve the synchronization robustness of coupled synthetic gene networks. The investigated robust synchronization criteria and proposed control methods are useful for a population of coupled synthetic gene networks with emergent synchronization behavior, especially for multi-cellular engineered gene networks in future.

References

Andrianantoandro, E., Basu, S., Karig, D.K. and Weiss, R. 2006. Synthetic biology: new engineering rules for an emerging discipline. Molecular Systems Biology 2: 1–14.

Arkin, A.P. and Fletcher, D.A. 2006. Fast, cheap and somewhat in control. Genome Biol 7: 114.

Atkinson, M.R., Savageau, M.A., Myers, J.T. and Ninfa, A.J. 2003. Development of genetic circuitry exhibiting toggle switch or oscillatory behavior in Escherichia coli. Cell 113: 597–607.

Ayala, F.J. 2007. Darwin's greatest discovery: design without designer. Proc Natl Acad Sci USA 104 Suppl 1: 8567–8573.

Basu, S., Mehreja, R., Thiberge, S., Chen, M.T. and Weiss, R. 2004. Spatiotemporal control of gene expression with pulse-generating networks. Proc Natl Acad Sci USA 101: 6355–6360.

Basu, S., Gerchman, Y., Collins, C.H., Arnold, F.H. and Weiss, R. 2005. A synthetic multicellular system for programmed pattern formation. Nature 434: 1130–1134.

Batt, G., Yordanov, B., Weiss, R. and Belta, C. 2007. Robustness analysis and tuning of synthetic gene networks. Bioinformatics 23: 2415.

Canton, B., Labno, A. and Endy, D. 2008. Refinement and standardization of synthetic biological parts and devices. Nature Biotechnology 26: 787–794.

Chen, B.S., Chang, C.H. and Lee, H.C. 2009. Robust synthetic biology design: stochastic game theory approach. Bioinformatics 25: 1822–1830.

Chen, B.S. and Wu, C.H. 2009. A systematic design method for robust synthetic biology to satisfy design specifications. BMC Syst Biol 3: 66.

Chen, B.S. and Chen, P.W. 2010. GA-based Design Algorithms for the Robust Synthetic Genetic Oscillators with Prescribed Amplitude, Period and Phase. Gene Regul Syst Bio 4: 35–52.

Chen, B.S. and Wu, C.H. 2010. Robust optimal reference-tracking design method for stochastic synthetic biology systems: T-S fuzzy approach. IEEE Transactions on Fuzzy Systems 18: 1144–1159.

Chen, B.S., Chang, C.H., Wang, Y.C., Wu, C.H. and Lee, H.C. 2011a. Robust model matching design methodology for a stochastic synthetic gene network. Math Biosci 230: 23–36.

Chen, B.S., Hsu, C.Y. and Liou, J.J. 2011b. Robust design of biological circuits: evolutionary systems biology approach. J Biomed Biotechnol 2011: 304236.

Chen, B.S. and Lin, Y.P. 2013. A Unifying Mathematical Framework for Genetic Robustness, Environmental Robustness, Network Robustness and their Trade-offs on Phenotype Robustness in Biological Networks. Part III: Synthetic Gene Networks in Synthetic Biology. Evol Bioinform Online 9: 87–109.

Chin, J.W. 2006. Programming and engineering biological networks. Current Opinion in Structural Biology 16: 551–556.

Dougherty, M.J. and Arnold, F.H. 2009. Directed evolution: new parts and optimized function. Current Opinion in Biotechnology 20: 486–491.

Ellis, T., Wang, X. and Collins, J.J. 2009. Diversity-based, model-guided construction of synthetic gene networks with predicted functions. Nat Biotechnol 27: 465–471.

Friedland, A.E., Lu, T.K., Wang, X., Shi, D., Church, G. and Collins, J.J. 2009. Synthetic gene networks that count. Science 324: 1199–1202.

Gardner, T.S., Cantor, C.R. and Collins, J.J. 2000. Construction of a genetic toggle switch in *Escherichia coli*. Nature 403: 339–342.

Gibbs, W.W. 2004. Synthetic life. Sci Am 290: 74–81.

Goh, K.I., Kahng, B. and Cho, K.H. 2008. Sustained oscillations in extended genetic oscillatory systems. Biophys J 94: 4270–4276.

Guido, N.J., Wang, X., Adalsteinsson, D., McMillen, D., Hasty, J., Cantor, C.R., Elston, T.C. and Collins, J.J. 2006. A bottom-up approach to gene regulation. Nature 439: 856–860.

Ham, T.S., Lee, S.K., Keasling, J.D. and Arkin, A.P. 2008. Design and construction of a double inversion recombination switch for heritable sequential genetic memory. PLoS One 3: e2815.

Haseltine, E.L. and Arnold, F.H. 2007. Synthetic gene circuits: design with directed evolution. Annual review of biophysics and biomolecular structure 36: 1–19.

Hasty, J., McMillen, D. and Collins, J.J. 2002. Engineered gene circuits. Nature 420: 224–230.

Hawkins, A.C., Arnold, F.H., Stuermer, R., Hauer, B. and Leadbetter, J.R. 2007. Directed evolution of Vibrio fischeri LuxR for improved response to butanoyl-homoserine lactone. Applied and environmental microbiology 73: 5775–5781.

Hooshangi, S., Thiberge, S. and Weiss, R. 2005. Ultra sensitivity and noise propagation in a synthetic transcriptional cascade. Proc Natl Acad Sci USA 102: 3581–3586.

Kobayashi, H., Kaern, M., Araki, M., Chung, K., Gardner, T.S., Cantor, C.R. and Collins, J.J. 2004. Programmable cells: interfacing natural and engineered gene networks. Proc Natl Acad Sci USA 101: 8414–8419.

Kramer, B.P., Viretta, A.U., Daoud-El-Baba, M., Aubel, D., Weber, W. and Fussenegger, M. 2004. An engineered epigenetic trans gene switch in mammalian cells. Nat Biotechnol 22: 867–870.

Lee, S.K., Chou, H., Ham, T.S., Lee, T.S. and Keasling, J.D. 2008. Metabolic engineering of microorganisms for biofuels production: from bugs to synthetic biology to fuels. Current Opinion in Biotechnology 19: 556–563.

Lu, T.K. and Collins, J.J. 2007. Dispersing biofilms with engineered enzymatic bacteriophage. Proceedings of the National Academy of Sciences 104: 11197–11202.

Lu, T.K. and Collins, J.J. 2009. Engineered bacteriophage targeting gene networks as adjuvants for antibiotic therapy. Proceedings of the National Academy of Sciences 106: 4629.

Pai, A., Tanouchi, Y., Collins, C.H. and You, L. 2009. Engineering multi cellular systems by cell-cell communication. Curr Opin Biotechnol 20: 461–470.

Rinaudo, K., Bleris, L., Maddamsetti, R., Subramanian, S., Weiss, R. and Benenson, Y. 2007. A universal RNAi-based logic evaluator that operates in mammalian cells. Nat Biotechnol 25: 795–801.

Simpson, M.L. 2006. Cell-free synthetic biology: a bottom-up approach to discovery by design. Mol Syst Biol 2: 69.

Sprinzak, D. and Elowitz, M.B. 2005. Reconstruction of genetic circuits. Nature 438: 443–448.

Steen, E.J., Chan, R., Prasad, N., Myers, S., Petzold, C.J., Redding, A., Ouellet, M. and Keasling, J.D. 2008. Metabolic engineering of Saccharomyces cerevisiae for the production of n-butanol. Microb Cell Fact 7: 36.

Stricker, J., Cookson, S., Bennett, M.R., Mather, W.H., Tsimring, L.S. and Hasty, J. 2008. A fast, robust and tunable synthetic gene oscillator. Nature 456: 516–519.

Tigges, M., Marquez-Lago, T.T., Stelling, J. and Fussenegger, M. 2009. A tunable synthetic mammalian oscillator. Nature 457: 309–312.

Waks, Z. and Silver, P.A. 2009. Engineering a synthetic dual-organism system for hydrogen production. Appl Environ Microbiol 75: 1867–1875.

Weber, W., Schoenmakers, R., Keller, B., Gitzinger, M., Grau, T., Daoud-El Baba, M., Sander, P. and Fussenegger, M. 2008. A synthetic mammalian gene circuit reveals antituberculosis compounds. Proceedings of the National Academy of Sciences 105: 9994–9998.

Win, M.N. and Smolke, C.D. 2007. A modular and extensible RNA-based gene-regulatory platform for engineering cellular function. Proc Natl Acad Sci USA 104: 14283–14288.

Win, M.N. and Smolke, C.D. 2008. Higher-order cellular information processing with synthetic RNA devices. Science 322: 456–460.

Withers, S.T. and Keasling, J.D. 2007. Biosynthesis and engineering of isoprenoid small molecules. Applied microbiology and biotechnology 73: 980–990.

Wu, C.H., Lee, H.C. and Chen, B.S. 2011a. Robust synthetic gene network design via library-based search method. Bioinformatics 27: 2700–2706.

Wu, C.H., Zhang, W. and Chen, B.S. 2011b. Multiobjective H_2/H_∞ synthetic gene network design based on promoter libraries. Math Biosci 233: 111–125.
Xie, Z., Wroblewska, L., Prochazka, L., Weiss, R. and Benenson, Y. 2011. Multi-input RNAi-based logic circuit for identification of specific cancer cells. Science 333: 1307–1311.
Yokobayashi, Y., Weiss, R. and Arnold, F.H. 2002. Directed evolution of a genetic circuit. Proceedings of the National Academy of Sciences of the United States of America 99: 16587–16591.

2

Mathematical Models and Design Specifications in Synthetic Gene Networks

2.1 Mathematical Models of Synthetic Gene Networks

First consider a simple cross-inhibition network shown in Figure 2.1. This network is synthesized with two genes, a and b, that code for two repressor proteins, A and B. More specifically, protein B represses the expression of gene a, whereas protein A represses the expression of gene b, and at higher concentration, the expression of its own gene. Protein degradations are not regulated. This synthetic system can be modeled by the following differential equations (Batt et al. 2007, Chen and Wu 2009).

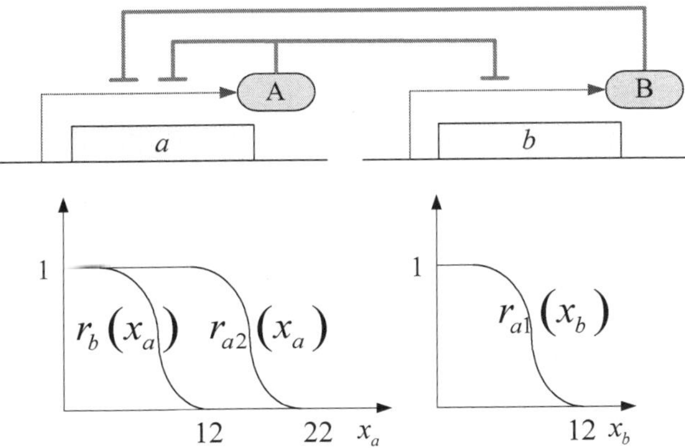

Figure 2.1. A single two-gene network.

$$\dot{x}_a = \kappa_a r_{a1}(x_b) r_{a2}(x_a) - \gamma_a x_a$$

$$\dot{x}_b = \kappa_b r_b(x_a) - \gamma_b x_b \tag{2.1}$$

The state variables x_a and x_b denote the concentrations of proteins A and B. κ's and γ's are the kinetic parameters and decay rates, respectively, and r's are the regulation functions, which capture the regulator effect of an effector protein on gene expression and are smooth sigmoidal functions (e.g., Hill functions) (de Jong 2002, Alon 2007).

We then consider the synthetic transcriptional cascade network shown in Figure 2.2. It consists of four genes: *tetR*, *lacI*, *cI*, and *eyfp* that code for three repressor proteins, TetR, LacI and CI, and the fluorescent protein EYFP, respectively. The fluorescence of the system, due to the protein EYFP, is the measured output. The protein CI inhibits gene *eyfp* and gene *tetR*. The protein TetR inhibits gene *lacI*. The protein LacI inhibits gene *cI*. The regulatory dynamic equations of the synthetic transcriptional cascade in Figure 2.2 are given as follows (Batt et al. 2007, Chen and Wu 2009).

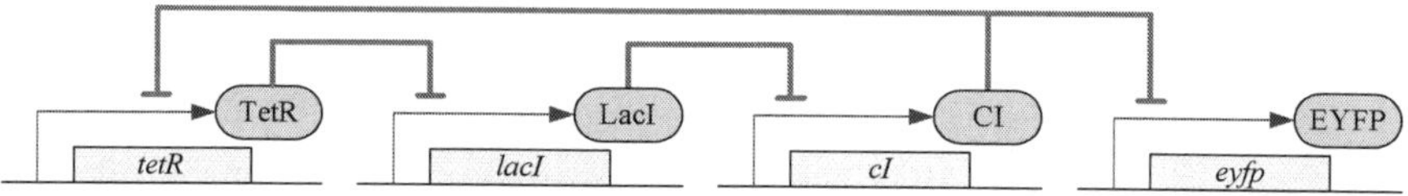

Figure 2.2. Synthetic transcriptional cascade network.

$$\dot{x}_{tetR} = \kappa_{tetR,0} + \kappa_{tetR} r_{tetR}(x_{cI}) - \gamma_{tetR} x_{tetR}$$

$$\dot{x}_{lacI} = \kappa_{lacI,0} + \kappa_{lacI} r_{lacI}(x_{tetR}) - \gamma_{lacI} x_{lacI}$$

$$\dot{x}_{cI} = \kappa_{cI,0} + \kappa_{cI} r_{cI}(x_{lacI}) - \gamma_{cI} x_{cI} \tag{2.2}$$

$$\dot{x}_{eyfp} = \kappa_{eyfp,0} + \kappa_{eyfp} r_{eyfp}(x_{cI}) - \gamma_{eyfp} x_{eyfp}$$

where $\kappa_{tetR,0}$, $\kappa_{lacI,0}$, $\kappa_{cI,0}$, and $\kappa_{eyfp,0}$ are the nominal generating ratios of the corresponding proteins. In addition, κ_{tetR}, κ_{lacI}, κ_{cI}, and κ_{eyfp} and γ_{tetR}, γ_{lacI}, γ_{cI}, and γ_{eyfp} are, respectively, the kinetic parameters and decay rates of the corresponding proteins. Furthermore, $r_{tetR}(x)$, $r_{lacI}(x)$, $r_{cI}(x)$, and $r_{eyfp}(x)$ are the Hill functions for repressors.

If a synthetic gene network consists of n genes, then the synthetic gene network of (2.1) or (2.2) can be extended to the following n-gene protein dynamics

$$\dot{x} = Nf(x) \tag{2.3}$$

where the state vector $x = \begin{bmatrix} x_1 \dots x_n \end{bmatrix}^T$ denotes the concentrations of proteins in the synthetic gene network. N denotes the corresponding stoichiometric matrix of the n-gene network.

In the above two synthetic gene networks, only the dynamic models of protein interaction are given. Actually, both the transcription and translation regulations are involved in synthetic gene network. In the repressilator shown in Figure 2.3, the first repressor protein, *lacI* from *E. coli*, inhibits the transcription of the second repressor gene, *tetR* from the tetracycline-resistance transposon *TN10*, whose protein product in turn inhibits the expression of the third gene, *cI* from the λ phage. Finally, *cI* inhibits *lacI* repression, completing the cycle. The negative feedback loop in the following transcriptional regulatory model can lead to temporal oscillations in the concentration of each component for us to design the repressilator and study its robust dynamic behavior (Elowitz and Leibler 2000, Chen and Chen 2010).

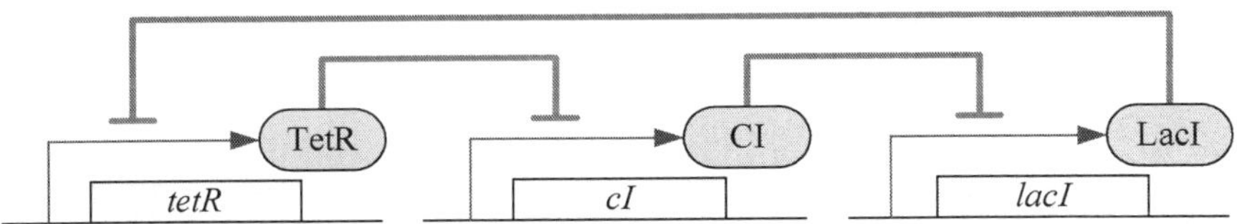

Figure 2.3. A repressilator network in *E. coli* host cell.

$$\frac{dm_i}{dt} = -\gamma_{m_i} m_i + \frac{\alpha_i}{1 + \left(p_j / K_i\right)^n} + \alpha_{0i}$$

$$\frac{dp_i}{dt} = \beta_i m_i - \gamma_{p_i} p_i$$

(2.4)

where m_i is the concentration of messenger RNA (mRNA) and p_i and p_j are concentrations of proteins for $i=1,2,3$ corresponding to LacI, TetR, and CI; and $j=3,1,2$ corresponding to CI, LacI, and TetR, respectively. Parameters γ_{mi} and γ_{pi} are the decay rates of mRNAs and proteins. α_i is the transcription rate of mRNA. α_{0i} is the effect of leakiness and is usually zero for stable state. β_i accounts for the number of translated protein molecules per mRNA molecule. K_i is the number of the jth proteins for a half repression of the ith promoter.

Another example is the biological AND gate shown in Figure 2.4. The biological AND gate generates an output signal only when it gets biochemical signals from both of its inputs. In the process, the input signal u_1 leads to the transcription of T7 polymerase gene, containing an early stop codon in the coding sequences that block translation. The input signal u_2 leads to the synthesis of a suppressor tRNA, which prevents the

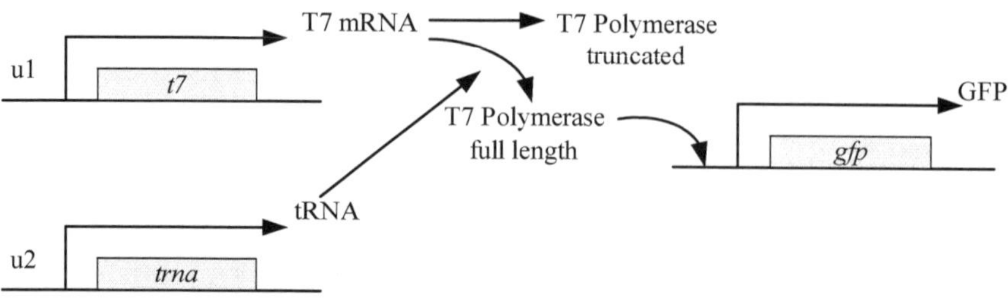

Figure 2.4. Synthetic biological AND gate.

premature termination and enables the translation of polymerase. When both of these inputs are present, the functional T7 RNA polymerase leads to the synthesis of the output signal, the reporter protein GFP. The following is the dynamic model of biological AND gate in Figure 2.4 (Terzer et al. 2007, Chen et al. 2011).

$$\frac{dx_{mT7Pol}}{dt} = k_1 u_1 - \lambda_1 x_{mT7Pol}$$

$$\frac{dx_{mtRNA}}{dt} = k_2 u_2 - \lambda_2 x_{mtRNA}$$

$$\frac{dx_{T7Pol^*}}{dt} = k_3 x_{mT7Pol} - \lambda_3 x_{T7Pol^*}$$

$$\frac{dx_{tRNA}}{dt} = k_4 x_{mtRNA} - \lambda_4 x_{tRNA}$$

$$\frac{dx_{T7Pol}}{dt} = k_5 x_{mT7Pol} x_{tRNA} - \lambda_5 x_{T7Pol}$$

$$\frac{dx_{mGFP}}{dt} = P_{mGFP} + \frac{k_6 \left(\dfrac{x_{T7Pol}}{K}\right)^n}{1 + \left(\dfrac{x_{T7Pol}}{K}\right)^n} - \lambda_6 x_{mGFP} \tag{2.5}$$

$$\frac{dx_{GFP}}{dt} = k_7 x_{mGFP} - \lambda_7 x_{GFP}$$

where x_{mT7Pol}, x_{mtRNA} and x_{mGFP} are the concentrations of mRNA transcribed from genes *T7Pol*, *tRNA* and *gfp*, respectively; concentrations of the corresponding proteins are represented by x_{T7Pol}, x_{tRNA} and x_{GFP}, respectively. k_1, k_2 and k_6 are the transcription rates. λ_1, λ_2 and λ_6 are the respective degradation rates of mRNA for *T7Pol*, *tRNA* and *gfp*. Parameters k_3, k_4 and k_7 are the translation rates of the proteins from the mRNAs, and λ_3,

λ_4 and λ_7 represent the degradation rates of protein non-functional T7 RNA polymerase, tRNA and GFP, respectively. x_{T7Pol} is the concentration of functional T7 RNA polymerase. k_5 is the reaction rate constant and λ_5 stands for the corresponding degradation constant. n is the Hill coefficient and K is the Hill constant. P_{mGFP} is the basal level.

Similarly, if a synthetic gene network consists of n genes, then the synthetic gene network in (2.4) or (2.5) can be extended to the following n-gene network dynamics

$$\dot{x} = f(x, k, u) \tag{2.6}$$

where $x = [\, x_{m1} \; x_{p1} \; \cdots \; x_{mn} \; x_{pn}]^T$ denotes the state vector of the n-gene synthetic network, $u = [u_1 \cdots u_m]^T$ denotes the input signal, $k = [k_1 \cdots k_l]^T$ denotes the kinetic parameters, $f(x, k, u)$ denotes the nonlinear biochemical interactions of synthetic gene network. The nonlinear dynamic model of synthetic gene network in (2.6) will be further discussed and designed in Chapter 4.

2.2 Design Specifications for Synthetic Gene Networks

The simple cross-inhibition network in (2.1) can be represented by the following stoichiometric matrix equation (Voit 2000, Chen and Wu 2009):

$$\begin{bmatrix} \dot{x}_a \\ \dot{x}_b \end{bmatrix} = \begin{bmatrix} \kappa_a & 0 & -\gamma_a & 0 \\ 0 & \kappa_b & 0 & -\gamma_b \end{bmatrix} \begin{bmatrix} r_{a1}(x_b) r_{a2}(x_a) \\ r_b(x_a) \\ x_a \\ x_b \end{bmatrix} \tag{2.7}$$

However, the stoichiometric matrix *in vivo* will suffer from the intrinsic parameter perturbations due to gene expression noises, splicing, mutation, evolution, etc. (McAdams and Arkin 1999, Chen and Wang 2006, Alon 2007) as

$$\begin{aligned} \kappa_a &\to \kappa_a + \Delta\kappa_a n_a & \gamma_a &\to \gamma_a + \Delta\gamma_a n_a \\ \kappa_b &\to \kappa_b + \Delta\kappa_b n_b & \gamma_b &\to \gamma_b + \Delta\gamma_b n_b \end{aligned} \tag{2.8}$$

where $\Delta\kappa_i$ and $\Delta\gamma_i$ denote the amplitudes of fluctuations of the stochastic kinetic parameters and decay rates; and n_i is a random white noise with zero mean and unit variance. Thus $\Delta\kappa_i$ and $\Delta\gamma_i$ denote the deterministic parts of parameter fluctuations and n_i absorbs the stochastic property of intrinsic parameter fluctuations. The independent variables n_a and n_b indicate that there are two independent stochastic sources of random parameter fluctuations. The covariance of stochastic intrinsic parameter fluctuation $\Delta\kappa_a n_a$ is given as $\mathrm{Cov}\big(\Delta\kappa_a n_a(t), \Delta\kappa_a n_a(\tau)\big) = \Delta\kappa_a^2 \delta_{t\tau}$, where $\delta_{t\tau}$ denotes the delta function, i.e., $\delta_{t\tau} = 1$ if $t = \tau$ and $\delta_{t\tau} = 0$ if $t \neq \tau$, i.e., $\Delta\kappa_i$ denotes the standard deviation σ_i of the stochastic parameter variation $\Delta\kappa_i n_i$.

Suppose the synthetic gene network also suffers from environmental disturbances due to changing extra-cellular environments and interactions with the cellular context in its host cell. Then the stochastic gene network can be represented as (Chen and Wu 2009).

$$
\begin{bmatrix} \dot{x}_a \\ \dot{x}_b \end{bmatrix} = \begin{bmatrix} \kappa_a + \Delta\kappa_a n_a & 0 & -\gamma_a - \Delta\gamma_a n_a & 0 \\ 0 & \kappa_b + \Delta\kappa_b n_b & 0 & -\gamma_b - \Delta\gamma_b n_b \end{bmatrix} \begin{bmatrix} r_{a1}(x_b) r_{a2}(x_a) \\ r_b(x_a) \\ x_a \\ x_b \end{bmatrix} + \begin{bmatrix} v_a \\ v_b \end{bmatrix}
$$

$$
= \begin{bmatrix} \kappa_a & 0 & -\gamma_a & 0 \\ 0 & \kappa_b & 0 & -\gamma_b \end{bmatrix} \begin{bmatrix} r_{a1}(x_b) r_{a2}(x_a) \\ r_b(x_a) \\ x_a \\ x_b \end{bmatrix} + \begin{bmatrix} \Delta\kappa_a & -\Delta\gamma_a \\ 0 & 0 \end{bmatrix} \begin{bmatrix} r_{a1}(x_b) r_{a2}(x_a) \\ x_a \end{bmatrix} n_a
$$

$$
+ \begin{bmatrix} 0 & 0 \\ \Delta\kappa_b & -\Delta\gamma_b \end{bmatrix} \begin{bmatrix} r_b(x_a) \\ x_b \end{bmatrix} n_b + \begin{bmatrix} v_a \\ v_b \end{bmatrix}
$$

$$
= N_2 f_2(x_2) + M_1 g_1(x_2) n_a + M_2 g_2(x_2) n_b + v_2 \tag{2.9}
$$

where $x_2 = [x_a \quad x_b]^T$ and $v_2 = [v_a \quad v_b]^T$ denote the state vector and the external disturbance of the synthetic gene network in the host cell, respectively. These intrinsic parameter fluctuations and external disturbances may cause the engineered gene network to be dysfunctional.

After employing the stochastic equation in (2.9) with intrinsic parameter fluctuations and external disturbances to mimic the realistic dynamic behaviors of the cross-inhibition network in the host cell, in order to work properly and efficiently, some design specifications for the synthetic gene network should be imposed as follows (Chen and Wu 2009).

i) The kinetic parameters and decay rates should be chosen from the following biologically feasible parametric ranges:

$$
\kappa_a \in [\kappa_{a1}, \kappa_{a2}], \gamma_a \in [\gamma_{a1}, \gamma_{a2}] \tag{2.10}
$$
$$
\kappa_b \in [\kappa_{b1}, \kappa_{b2}], \gamma_b \in [\gamma_{b1}, \gamma_{b2}]
$$

ii) The intrinsic stochastic parameter fluctuations with the following standard deviations must be tolerated,

$$
\Delta\kappa_a, \Delta\gamma_a, \Delta\kappa_b, \Delta\gamma_b, \tag{2.11}
$$

which are requested by designers to meet the *in vivo* conditions in the host cell.

iii) The following desired steady states must be achieved to meet some bio-design purposes:

$$x_a \rightarrow x_{ad}, \; x_b \rightarrow x_{bd} \text{ at } t \rightarrow \infty \tag{2.12}$$

iv) The external disturbances must be attenuated to remain below a prescribed attenuation level ρ, i.e., the effect of external disturbances on the regulation error around the desired steady state in (iii) should be less than ρ^2 from the mean energy point of view

$$\frac{E\int_0^\infty (x_a - x_{ad})^2 + (x_b - x_{bd})^2 \, dt}{E\int_0^\infty (v_a^2 + v_b^2) \, dt} < \rho^2 \tag{2.13}$$

for all possible bounded disturbances v_a and v_b. This is also a design specification for the noise filtering ability of the synthetic gene network, i.e., with a filtering ability of ρ to attenuate the external disturbances v_a and v_b (Zhang et al. 2005, Zhang and Chen 2006). In (2.13), we do not need to know the statistics of external disturbances v_a and v_b but are concerned only with the attenuation level (i.e., the ratio ρ) of external disturbances. If v_a and v_b are deterministic signals, the expectation E on v_a and v_b can be neglected.

Our design goal is to choose two kinetic parameters κ_a and κ_b and two decay rates γ_a and γ_b from the feasible parameter ranges in (2.10) so that the desired steady states x_{ad} and x_{bd} in (2.12) can be achieved under the above specified stochastic parameter variations and stochastic external disturbances, i.e., the allowable standard deviations of stochastic parameter fluctuations in (2.11) should be tolerated and the external disturbances should be attenuated below a prescribed attenuation level ρ in (2.13). If the above four design specifications (i)–(iv) can be imposed in the design procedure of the synthetic gene network, then the engineered synthetic gene network could work properly and efficiently in the host cell under intrinsic parameter fluctuations and external disturbances.

Now, we consider the n-gene synthetic network case in (2.3). Suppose the n-gene synthetic network suffers from the intrinsic parameter perturbations and the environmental disturbances, then the dynamic model in (2.3) can be rewritten as the following stochastic dynamic system (Chen and Wu 2009).

$$\dot{x} = Nf(x) + \sum_{i=1}^{m} M_i g_i(x) n_i + v \tag{2.14}$$

where the state vector $x = [x_1 \dots x_n]^T$ denotes the concentrations of proteins in the synthetic gene network. N denotes the corresponding stoichiometric matrix of the n-gene network. $M_i, \; i = 1, \dots, m$, denotes the fluctuation matrices due to independent random noise sources n_i, $i = 1,\ldots,m$, and the elements of M_i denote the standard deviations of the corresponding parameter fluctuations. $v = [v_1 \ldots v_n]^T$ denotes the vector of external disturbances. The stochastic system in (2.14) is used to mimic the realistic dynamic behavior of a synthetic gene network of n genes in the host cell. This network, however, suffers from the intrinsic parameter fluctuations and external disturbances in the context of the host cell. Thus, a robust synthetic gene network should be designed with the ability not only to tolerate these parameter fluctuations and attenuate the external disturbances from the environments but also to achieve the desired steady state behaviors.

For convenience of analysis and design, the stochastic dynamic equation (2.14) of a more general stochastic gene network can be represented by the following Ito's stochastic differential equation (Chen and Hsu 1995, Zhang et al. 2005, Zhang and Chen 2006)

$$dx = \left(Nf(x) + v\right)dt + \sum_{i=1}^{m} M_i g_i(x)dW_i \tag{2.15}$$

where $W_i(t)$ is a standard Wiener process with $dW_i(t) = n_i(t)dt$.

The design specifications in (2.10)–(2.13) can also be extended as follows for a more general synthetic gene network in (2.14) (Chen and Wu 2009):

i) The kinetic parameters and the decay rates in stoichiometric matrix should be chosen from the following biologically feasible range
$$N \in [N_1, N_2] \tag{2.16}$$

ii) The stochastic kinetic parameters and decay rate fluctuations with prescribed standard deviations in M_i in the following state-dependent noise terms

$$\sum_{i=1}^{m} M_i g_i(x)dW_i \tag{2.17}$$

should be tolerated by the synthetic gene network.

iii) The following desired steady state should be achieved
$$x \to x_d \text{ as } t \to \infty \tag{2.18}$$

where x_d is the desired steady state specified by the designer for some design purposes of the synthetic gene network.

iv) The following prescribed disturbance filtering ability (i.e., the H_∞ filtering) should be achieved (Zhang et al. 2005)

$$\frac{E\int_0^\infty (x - x_d)^T Q(x - x_d) \; dt}{E\int_0^\infty v^T v dt} < \rho^2 \tag{2.19}$$

for all bounded $v(t)$, where $Q \geq 0$ is a symmetric weighting matrix and ρ is a prescribed attenuation level less than 1. That is, the effect of external disturbance v on the regulation error $x - x_d$ should be less than the attenuation level ρ from the average energy perspective. In this situation, the synthetic gene network can efficiently attenuate the effect of external disturbances on the regulation of system state x to the desired steady state x_d. The design specification of a prescribed attenuation level ρ may be a trade off between the filtering ability and the specification (i) in (2.16), i.e., a small ρ (i.e., a strict specification of attenuation level) may lead to a small feasible range of N, which may be outside the allowable range in (2.16). In (2.19), if the external disturbance is deterministic, then the expectation E can be neglected.

References

Alon, U. 2007. An Introduction to Systems Biology: Design Principles of Biological Circuits. Chapman & Hall/CRC, London.

Batt, G., Yordanov, B., Weiss, R. and Belta, C. 2007. Robustness analysis and tuning of synthetic gene networks. Bioinformatics 23: 2415.

Chen, B.S. and Wang, Y.C. 2006. On the attenuation and amplification of molecular noise in genetic regulatory networks. BMC Bioinformatics 7: 52.

Chen, B.S. and Wu, C.H. 2009. A systematic design method for robust synthetic biology to satisfy design specifications. BMC Syst Biol 3: 66.

Chen, B.S. and Chen, P.W. 2010. GA-based Design Algorithms for the Robust Synthetic Genetic Oscillators with Prescribed Amplitude, Period and Phase. Gene Regul Syst Bio 4: 35-52.

Chen, B.S., Hsu, C.Y. and Liou, J.J. 2011. Robust design of biological circuits: evolutionary systems biology approach. J Biomed Biotechnol 2011: 304236.

Chen, G. and Hsu, S.-H. 1995. Linear Stochastic Control Systems. Boca Raton, FL: CRC Press.

de Jong, H. 2002. Modeling and simulation of genetic regulatory systems: a literature review. Journal of Computational Biology 9: 67–103.

Elowitz, M.B. and Leibler, S. 2000. A synthetic oscillatory network of transcriptional regulators. Nature 403: 335–338.

McAdams, H.H. and Arkin, A. 1999. It's a noisy business! Genetic regulation at the nanomolar scale. Trends Genet 15: 65–69.

Terzer, M., Jovanovic, M., Choutko, A., Nikolayeva, O., Korn, A., Brockhoff, D., Zurcher, F., Friedmann, M., Schutz, R., Zitzler, E., Stelling, J. and Panke, S. 2007. Design of a biological half adder. Synthetic Biology, IET 1: 53–58.

Voit, E.O. 2000. Computational Analysis of Biochemical Systems: A Practical Guide for Biochemists and Molecular Biologists. Cambridge University Press, Cambridge.

Zhang, W., Chen, B.S. and Tseng, C.S. 2005. Robust H∞ filtering for nonlinear stochastic systems. IEEE Transactions on Signal Processing 53: 589–598.

Zhang, W. and Chen, B.S. 2006. State feedback H∞ control for a class of nonlinear stochastic systems. SIAM Journal on Control and Optimization 44: 1973–1991.

3

Robust Synthetic Biology Designs based on System Dynamic Models

3.1 Robust Synthetic Biology Design to Satisfy Design Specifications

Based on the stochastic gene network in (2.15) and the design specifications (2.16)–(2.19) in previous chapter, we now can introduce the robust synthetic biology design to satisfy design specifications. Our design goal is to choose some kinetic parameters and decay rates in the stoichiometric matrix N from the biologically feasible parameter range $[N_1, N_2]$ such that the desired steady state x_d in (2.18) can be achieved, the stochastic parameter variations $\sum_{i=1}^{m} M_i g_i(x) dW_i$ can be robustly tolerated (stabilized), and the prescribed disturbance attenuation level ρ on v in (2.19) can be achieved.

Based on the analyses in the above, the design problem of robust synthetic gene networks becomes how to specify the kinetic parameters and decay rates in the stoichiometric matrix N in (2.15) such that the design specifications (2.16)–(2.19) must be satisfied to let the synthetic gene network work properly *in vivo* under intrinsic parameter fluctuations and external disturbances. In order to achieve the desired steady state x_d, for the convenience of design, the origin of the nonlinear stochastic system in (2.15) should be shifted to x_d. In such a situation, if the shifted nonlinear stochastic system is stabilized at the origin, then the desired steady state x_d will be equivalently achieved. This will simplify the design procedure. Let us denote $\tilde{x} = x - x_d$, then we get the following shifted stochastic system (Slotine and Li 1991, Chen and Wu 2009).

$$d\tilde{x} = \left(Nf\left(\tilde{x}+x_d\right)+v\right)dt + \sum_{i=1}^{m} M_i g_i\left(\tilde{x}+x_d\right)dW_i \tag{3.1}$$

i.e., the origin $\tilde{x} = 0$ of stochastic system in (3.1) is at the desired steady state x_d of the original stochastic system in (2.15).

For the stochastic system in (3.1), if we specify $N \in [N_1, N_2]$ such that the origin $\tilde{x} = 0$ can be robustly stabilized to tolerate the stochastic parameter fluctuation $\sum_{i=1}^{m} M_i g_i\left(\tilde{x}+x_d\right)dW_i$ and efficiently attenuate the external disturbance v to the following prescribed level (i.e., H_∞ filtering ability)

$$\frac{E\int_0^\infty \tilde{x}^T Q\tilde{x}dt}{E\int_0^\infty v^T vdt} < \rho^2 \text{ or } E\int_0^\infty \tilde{x}^T Q\tilde{x}dt < \rho^2 E\int_0^\infty v^T vdt \tag{3.2}$$

then the design specifications (2.16)–(2.19) can be achieved for the stochastic gene network in (2.15) simultaneously under intrinsic parameter fluctuations and external disturbances in the host cell. If the initial condition is also considered (Boyd et al. 1994, Chen and Zhang 2004, Chen et al. 2008, Chen and Wu 2009), then the filtering ability in the inequality (3.2) should be modified as

$$E\int_0^\infty \tilde{x}^T Q\tilde{x}dt \leq V\left(\tilde{x}(0)\right)+\rho^2 E\int_0^\infty v^T vdt, \; \forall \tilde{x}(0) \neq 0 \tag{3.3}$$

for some positive function $V(\tilde{x})$.

According to the above analyses, we can design kinetic parameters and decay rates in $N \in [N_1, N_2]$ of the stochastic gene network in (3.1) to achieve both the robust stabilization to tolerate the stochastic parameter fluctuation and the filtering ability of external disturbance in (3.3). This is called the robust synthetic gene network design problem (Chen et al. 2008, Chen and Wu 2009). Before further analysis of the robust stabilization and filtering design problem of stochastic synthetic gene networks, we first consider the robust stabilization to tolerate intrinsic stochastic parameter fluctuation in (3.1) in the case free from external disturbance (i.e., $v \equiv 0$). From the theory of stochastic stability, the stochastic synthetic gene network in (3.1) with $v(t) = 0$ is assumed with asymptotic stability in probability if the expectation of the time derivative of Lyapunov (energy) function $V(\tilde{x})$

is negative (Chen and Hsu 1995, Chen and Zhang 2004, Chen et al. 2008, Chen and Wu 2009), i.e.,

$$E\left(\frac{d}{dt}V(\tilde{x})\right)<0 \tag{3.4}$$

where $V(\tilde{x})>0$ is the Lyapunov (energy) function of the synthetic gene network in (3.1). The inequality in (3.4) means that on average the energy function of the synthetic gene network decreases with time and will asymptotically converge to $\tilde{x}=0$ or $x\to x_d$ in probability in the case of $v(t)=0$. In the case $v(t)\neq 0$, only the H_∞ disturbance attenuation level in (3.2) or (3.3) can be designed because the asymptotical stability in probability cannot be achieved due to the continuous interference of external disturbances, i.e., $\tilde{x}\to 0$ or $x\to x_d$ cannot be achieved as $t\to\infty$ and the deviation from x_d (i.e., $\tilde{x}$) due to external disturbances can only be attenuated to a level ρ by the design specification of noise filtering ability in (3.2) or (3.3).

From the stochastic network in (3.1), we obtain the following result:

Proposition 3.1: If some design kinetic parameters and decay rates in $N\in[N_1,N_2]$ are chosen such that the following Hamilton-Jacobi inequality (HJI) has a positive solution $V(\tilde{x})>0$

$$\left(\frac{\partial V(\tilde{x})}{\partial\tilde{x}}\right)^T Nf(\tilde{x}+x_d)+\tilde{x}^T Q\tilde{x}+\frac{1}{4\rho^2}\left(\frac{\partial V(\tilde{x})}{\partial\tilde{x}}\right)^T\left(\frac{\partial V(\tilde{x})}{\partial\tilde{x}}\right)$$
$$+\frac{1}{2}\sum_{i=1}^{m}g_i^T(\tilde{x}+x_d)M_i^T\frac{\partial^2 V(\tilde{x})}{\partial\tilde{x}^2}M_i g_i(\tilde{x}+x_d)<0 \tag{3.5}$$

then (a) the stochastic gene network in (3.1) can achieve both the robust stabilization to tolerate intrinsic stochastic parameter perturbations and the prescribed attenuation level ρ on the external disturbances, i.e., the design specifications (i), (ii) and (iv) in (2.16), (2.17) and (2.19), respectively, are all satisfied; (b) if the stochastic gene network is free of external disturbances, i.e., $v(t)=0$, then the shifted gene network in (3.1) will asymptotically converge to $\tilde{x}=0$ or $x\to x_d$ in probability, or equivalently, the original stochastic gene network in (2.15) will asymptotically converge to the desired steady state x_d in probability, i.e., the design specification (iii) in (2.18) is achieved.

Proof: See Appendix 3.1

Remark 3.1: If the synthetic gene network is free of external disturbances and only the stochastic parameter fluctuations are to be robustly tolerated, the HJI in (3.5) is reduced to the following inequality

$$\frac{\partial V(\tilde{x})}{\partial \tilde{x}}^{T} Nf(\tilde{x}+x_d) + \tilde{x}^T Q\tilde{x} + \frac{1}{2}\sum_{i=1}^{m} g_i^T(\tilde{x}+x_d) M_i^T \frac{\partial^2 V(\tilde{x})}{\partial \tilde{x}^2} M_i g_i(\tilde{x}+x_d) < 0 \quad (3.6)$$

without the disturbance attenuation-related term $\dfrac{1}{4\rho^2}\left(\dfrac{\partial V(\tilde{x})}{\partial \tilde{x}}\right)^T \dfrac{\partial V(\tilde{x})}{\partial \tilde{x}}$.

It is easier to find a positive solution $V(\tilde{x}) > 0$ to satisfy the HJI in (3.6) than in (3.5). Furthermore, the synthetic gene network design which satisfies (3.6) could achieve asymptotical convergence in probability to the desired steady states in the disturbance free case.

In the above discussion, we only focus on the parameter perturbations which are allowed in the stoichiometric matrix of the nonlinear model and environment. Suppose the perturbations are also allowed in nonlinear functions governing the synthetic biological system, i.e., r-function in (2.1) also suffers from the stochastic perturbations $r_i(t) \rightarrow r_i(t) + \Delta r_i(t)$ such that $f(x)$ and $g_i(x)$ of the nonlinear genetic system in (3.1) suffer from the stochastic perturbations $f(x) \rightarrow f(x) + \Delta f(x)$ and $g_i(x) \rightarrow g_i(x) + \Delta g_i(x)$, respectively. In this situation, the nonlinear synthetic gene network suffers from the following parametric and functional perturbations (Chen and Wu 2009).

$$d\tilde{x} = \left(N\left(f(\tilde{x}+x_d) + \Delta f(\tilde{x}+x_d)\right) + v\right)dt$$
$$+\sum_{i=1}^{m} M_i\left(g_i(\tilde{x}+x_d) + \Delta g_i(\tilde{x}+x_d)\right)dW_i \quad (3.7)$$

Suppose the functional perturbations are bounded by the following sectors, $E\|\Delta f(x)\|_2 \leq \alpha E\|x\|_2$, $E\|\Delta g_i(x)\|_2 \leq \beta_i E\|x\|_2$ or equivalently,

$$E\left(\Delta f(x)\right)^T \left(\Delta f(x)\right) \leq \alpha^2 Ex^T x$$
$$E\left(\Delta g_i(x)\right)^T \left(\Delta g_i(x)\right) \leq \beta_i^2 Ex^T x \quad (3.8)$$

Then we can obtain the following result:

Proposition 3.2: Suppose the synthetic gene network suffers from the parametric variations and functional perturbations as (3.7) and (3.8). If some design kinetic parameters and decay rates in $N \in [N_1, N_2]$ are chosen such that the following HJI has a positive solution $V(\tilde{x}) > 0$

$$\left(\frac{\partial V(\tilde{x})}{\partial \tilde{x}}\right)^T Nf(\tilde{x}+x_d) + \tilde{x}^T Q\tilde{x} + \frac{1}{2}\sum_{i=1}^{m} g_i^T(\tilde{x}+x_d)M_i^T \frac{\partial^2 V(\tilde{x})}{\partial \tilde{x}^2}M_i g_i(\tilde{x}+x_d)$$

$$+ \frac{1}{4\rho^2}\left(\frac{\partial V(\tilde{x})}{\partial \tilde{x}}\right)^T\left(\frac{\partial V(\tilde{x})}{\partial \tilde{x}}\right) + \frac{1}{4}\left(\frac{\partial V(\tilde{x})}{\partial \tilde{x}}\right)^T NN^T \frac{\partial V(\tilde{x})}{\partial \tilde{x}} \qquad (3.9)$$

$$+ \alpha^2(\tilde{x}+x_d)^T(\tilde{x}+x_d) + \frac{1}{2}\sum_{i=1}^{m}\beta_i^T(\tilde{x}+x_d)M_i^T \frac{\partial^2 V(\tilde{x})}{\partial \tilde{x}^2}M_i g_i(\tilde{x}+x_d) < 0$$

then there are two results: (a) the stochastic gene network in (3.7) can achieve H_∞ robust stabilization to tolerate parametric variations and functional perturbation, and can reach the prescribed disturbance filtering ability ρ to attenuate the external disturbances; and (b) if the stochastic gene network in (3.7) and (3.8) is free of external disturbances, i.e., $v(t) = 0$, it will asymptotically converge to $\tilde{x} = 0$ or $x \rightarrow x_d$ in probability.

Proof: See Appendix 3.2

Remark 3.2: Comparing Proposition 3.1 and Proposition 3.2, it is seen that there are three extra terms in (3.9) due to the stochastic function perturbations. It is more difficult to find design parameters in $N \in [N_1, N_2]$ to solve $V(\tilde{x}) > 0$ for HJI in (3.9) than to find parameters for HJI in (3.5) because the stochastic gene system in (3.7) has to tolerate not only the stochastic parameter variations but also the functional perturbations.

In general, it is very difficult to specify $N \in [N_1, N_2]$ to solve HJI in (3.5), (3.6) or (3.9) for $V(\tilde{x}) > 0$ via the systematic method. At present, there is no good method to solve the nonlinear partial differential HJI analytically or numerically. In this situation, the global linearization technique is employed to transform the nonlinear stochastic gene network in (3.1) to an interpolation of a set of globally linearized gene networks to simplify the design procedure. By the global linearization method (Boyd et al. 1994, Chen and Wu 2009), if all the global linearizations are bound by a polytope consisting of M vertices as

$$
\begin{pmatrix} \dfrac{\partial f\left(\tilde{x}+x_d\right)}{\partial \tilde{x}} \\[2mm] \dfrac{\partial g_1\left(\tilde{x}+x_d\right)}{\partial \tilde{x}} \\[2mm] \vdots \\[2mm] \dfrac{\partial g_m\left(\tilde{x}+x_d\right)}{\partial \tilde{x}} \end{pmatrix} \in \mathrm{Co}\left(\begin{pmatrix} F_1 \\ G_{11} \\ \vdots \\ G_{m1} \end{pmatrix}, \cdots, \begin{pmatrix} F_M \\ G_{1M} \\ \vdots \\ G_{mM} \end{pmatrix} \right), \forall \tilde{x} \tag{3.10}
$$

where Co denotes the convex hull of polytope with M vertices defined in (3.10), then the state trajectories $\tilde{x}(t)$ of the shifted gene network in (3.1) will belong to the convex combination of the state trajectories of the following M linearized synthetic gene networks derived from the vertices of the polytope in (3.10) (Boyd et al. 1994, Chen and Wu 2009)

$$
d\tilde{x} = \left(NF_j\tilde{x}+v\right)dt + \sum_{i=1}^{m} M_iG_{ij}\tilde{x}dW_iE, j = 1,\cdots, M \tag{3.11}
$$

By the global linearization theory (Boyd et al. 1994), if (3.10) holds, then every trajectory of the nonlinear synthetic gene network in (3.1) is a trajectory of a convex combination of M linearized synthetic gene networks in (3.11). Therefore, if we can prove that the convex combination of M linearized synthetic gene networks in (3.11) can tolerate the intrinsic parameter fluctuations and attenuate the external disturbances below a prescribed level, then the original nonlinear synthetic gene network in (3.1) will have the same robust stabilization and disturbance attenuation property. The convex combination of M linearized gene networks in (3.11) can be written as

$$
d\tilde{x} = \sum_{j=1}^{M}\alpha_j\left(\tilde{x}\right)\left(NF_j\tilde{x}dt + \sum_{i=1}^{m}M_iG_{ij}\tilde{x}dW_i \right) + vdt \tag{3.12}
$$

where the interpolation function $\alpha_j(\tilde{x})$ satisfies $0 \le \alpha_j\left(\tilde{x}\right)\le 1$ and $\sum_{j=1}^{M}\alpha_j\left(\tilde{x}\right)=1$

, i.e., the trajectory of nonlinear synthetic gene network in (3.1) could be

represented by the interpolated synthetic gene network in (3.12), which is the convex combination of M linearized gene networks in (3.11). Therefore, the following result can be obtained.

Proposition 3.3: Assume that some design kinetic parameters and decay rates in $N \in [N_1, N_2]$ are chosen such that the following M inequalities have a common symmetric positive definite solution $P > 0$

$$PNF_j + F_j^T N^T P + \sum_{i=1}^{m} G_{ij}^T M_i^T PM_i G_{ij} + Q + \frac{1}{\rho^2} PP^T < 0 \tag{3.13}$$

then there are two results: (a) the stochastic gene network with parametric variations and external disturbances in (3.1) will be robustly stable to tolerate intrinsic stochastic parameter perturbation and also achieve a prescribed attenuation level ρ on the external disturbance, i.e., the design specifications (i), (ii) and (iv) in (2.16), (2.17) and (2.19) are all satisfied; and (b) if the gene network is free of external disturbance, i.e., $v(t) = 0$, then the gene network in (3.1) will asymptotically converge to $\tilde{x} = 0$ in probability, or equivalently, the original synthetic gene network in (2.15) will asymptotically converge to the desired steady state x_d in probability, i.e., the design specification (iii) in (2.18) is achieved.

Proof: See Appendix 3.3.

Similarly, for the stochastic gene network in (3.7) with parameter variations, functional perturbations and noises, based on global linearization method, we obtain the following result for Proposition 3.2.

Proposition 3.4: Assume some design kinetic parameters and decay rates in $N \in [N_1, N_2]$ are chosen such that the following M inequalities have a common symmetric positive definite solution $P > 0$

$$PNF_j + F_j^T N^T P + PNN^T P + \alpha^2 I + Q$$

$$+ \sum_{i=1}^{m} \left(\beta_i^2 M_i^T PM_i + G_{ij}^T M_i^T PM_i G_{ij} \right) + \frac{1}{\rho^2} PP^T < 0, \ j = 1,\ldots,M \tag{3.14}$$

then there are two results: (a) the synthetic gene network with parameter variations, functional perturbations and external disturbances in (3.7) will be robustly stable to tolerate intrinsic parameter variation and functional perturbations, and achieve a prescribed attenuation level ρ on the external disturbances; and (b) if the synthetic gene network is free of external disturbances, then the synthetic gene network in (3.7) will asymptotically converge to $\tilde{x} = 0$ in probability, or $x \rightarrow x_d$ asymptotically in probability.

Proof: Similar to Proposition 3.3

Remark 3.3:

i) By Schur complement (Boyd et al. 1994), the inequalities in (3.13) could be transformed to the following linear matrix inequalities (LMIs)

$$\begin{bmatrix} PNF_j + F_j^T N^T P + \sum_{i=1}^{m} G_{ij}^T M_i^T PM_i G_{ij} + Q & P \\ P & -\rho^2 I \end{bmatrix} < 0, \; j = 1, \ldots, M \quad (3.15)$$

The robust synthetic gene network design problem by specifying $N \in [N_1, N_2]$ to solve a positive function $V(\tilde{x}) > 0$ for HJI in (3.5) with a prescribed disturbance attenuation level ρ is transformed into the problem of specifying $N \in [N_1, N_2]$ to solve a common positive symmetric definite matrix $P > 0$ for a set of inequalities in (3.13), or equivalently for a set of LMIs in (3.15). The LMIs in (3.15) can be efficiently solved by the so-called interior-point method (Boyd et al. 1994). It has been proven that the computational complexity for solving LMIs in (3.15) via the interior point method for the n-gene network in (2.14) is about the order $O\left(m^{2.75} M^{1.5}\right)$ of arithmetic operations, where

$m = \dfrac{1}{2} n(n+1)$ and M is the number of linearized systems (Boyd et al.

1994). The LMIs in (3.15) could be efficiently solved by the LMI toolbox in Matlab (Gahinet et al. 1995, Chen and Wu 2009).

Similarly, by Schur complement (Boyd et al. 1994), the inequalities in (3.14) are equivalent to specifying $N \in [N_1, N_2]$ to solve $P > 0$ for the following LMIs

$$\begin{bmatrix} \begin{aligned} & PNF_j + F_j^T N^T P + \alpha^2 I + Q \\ & + \sum_{i=1}^{m} \left(\rho_i^2 M_i^T PM_i + G_{ij}^T M_i^T PM_i G_{ij} \right) \end{aligned} & P \\ P & -\left(NN^T + \dfrac{1}{\rho^2} I \right)^{-1} \end{bmatrix} < 0 \quad (3.16)$$

ii) If the synthetic gene network is free of external disturbances and the robust stabilization only needs to tolerate the stochastic parameter fluctuation, then the inequalities in (3.13) will be reduced to the following LMIs

$$PNF_j + F_j^T N^T P + Q + \sum_{i=1}^{m} G_{ij}^T M_i^T PM_i G_{ij} < 0, \ j = 1, \cdots, M \qquad (3.17)$$

without the term $\dfrac{1}{\rho^2} PP^T$ in (3.13). In this situation, it is easier to specify the kinetic parameters and decay rates in N to satisfy the above LMIs. Furthermore, the asymptotic convergence to the desired steady states x_d in probability can also be achieved.

iii) In addition to the global linearization method in this section, a piecewise-affine model for nonlinear gene regulatory network has also been introduced to consider geometric constraints of genetic regulatory network (Drulhe et al. 2008).

iv) Using global linearization, every trajectory of a nonlinear system in (3.1) is also a trajectory of the convex combinatory system in (3.11). However, there are many trajectories of the convex combinatory system that are not trajectories of the nonlinear system (Boyd et al. 1994). Therefore, the conditions of Proposition 3.3 are more constraining than the ones of Proposition 3.1. Hence, the solution of Proposition 3.3 is more conservative than the one of Proposition 3.1. Similarly, the solution of Proposition 3.4 is more conservative than the one of Proposition 3.2 because the conditions of Proposition 3.4 are more constraining than the ones of Proposition 3.2.

Based on the above analyses, the design problem of robust synthetic gene network becomes how to select an adequate N from the allowable range $[N_1, N_2]$ to satisfy the LMIs in (3.15) to meet the design specifications (i)–(iv) in (2.16)–(2.19). In order to simplify the selection process of N, we define

$$N_0 = \frac{1}{2}(N_2 + N_1), \ \tilde{N} = \frac{1}{2}(N_2 - N_1) \qquad (3.18)$$

where N_0 denotes the nominal value and $\tilde{N}$ denotes the allowable range from the nominal value. Let

$$N = N_0 + \Delta N, \text{ where } \Delta N \in \left[-\tilde{N}, \tilde{N} \right] \qquad (3.19)$$

i.e., we could select the nominal N_0 for N at first and then add a fine tuning $\Delta N \in \left[-\tilde{N}, \tilde{N} \right]$ around the nominal N_0 to meet LMIs in (3.15) or we could select fine tuning $\Delta N \in \left[-\tilde{N}, \tilde{N} \right]$ to meet the following LMIs to simplify the design procedure.

$$\left[\begin{array}{cc} P\left(N_0 + \Delta N\right)F_j + F_j^T\left(N_0 + \Delta N\right)P + \displaystyle\sum_{i=1}^{m} G_{ij} M_i^T P M_i G_{ij} + Q & P \\ P & -\rho^2 I \end{array} \right] < 0,\ j = 1,2,\cdots,M \quad (3.20)$$

Then developing the robust synthetic network requires finding a fine tuning ΔN from the allowable range $\left[-\tilde{N}, \tilde{N}\right]$ to meet a positive matrix $P > 0$ solution of LMIs in (3.20), which can be achieved via the help of the LMI toolbox in Matlab. The detailed search process for fine tuning ΔN is given in the design example in the sequel.

From the analyses above, a design procedure for a robust synthetic gene network is proposed as follows (Chen and Wu 2009):

1) Provide the design specification of robust synthetic gene network in (2.16)–(2.19).
2) Shift the desired steady state x_d to the origin, as in (3.1).
3) Perform the global linearization as in (3.10) to obtain F_i and G_{ij}.
4) Find the nominal $N_0 = \dfrac{1}{2}\left(N_2 + N_1\right)$ and solve LMIs for fine tuning ΔN from the allowable range $\left(-\tilde{N}, \tilde{N}\right)$.
5) Find the design kinetic parameters and decay rates of the synthetic gene network as $N = N_0 + \Delta N$.

3.2 An *in silico* Design Example

Let us consider the synthetic transcriptional cascade network in Figure 2.2. In the dynamic equations of the synthetic transcriptional cascade network (2.2), the nominal generating ratios of the corresponding proteins, $\kappa_{tetR,0}$, $\kappa_{lacI,0}$, $\kappa_{cI,0}$, and $\kappa_{eyfp,0}$, are assumed to be 150, 587, 210, and 3487, respectively, but with stochastic parameter fluctuations. In addition, The Hill function is a decreasing S-shaped curve, which can be described in the form

$$r_i(x) = \frac{\beta}{1 + \left(\dfrac{x}{K_i}\right)^n} \quad \text{with } \beta = 1,\ n = 2,\ K_i = 1000,\ i = tetR, lacI, cI, eyfp$$

(Alon 2007).

According to the stochastic gene network in (2.15), the stochastic gene network with four random parameter fluctuation sources in (2.2) can be represented by

$$
\begin{aligned}
\begin{bmatrix} dx_{tetR} \\ dx_{lacI} \\ dx_{cI} \\ dx_{eyfp} \end{bmatrix} =
& \left(\begin{bmatrix}
\kappa_{tetR,0} & -\gamma_{tetR} & 0 & 0 & 0 & \kappa_{tetR} & 0 & 0 & 0 \\
\kappa_{lacI,0} & 0 & -\gamma_{lacI} & 0 & 0 & 0 & \kappa_{lacI} & 0 & 0 \\
\kappa_{cI,0} & 0 & 0 & -\gamma_{cI} & 0 & 0 & 0 & \kappa_{cI} & 0 \\
\kappa_{eyfp,0} & 0 & 0 & 0 & -\gamma_{eyfp} & 0 & 0 & 0 & \kappa_{eyfp}
\end{bmatrix}
\begin{bmatrix} 1 \\ x_{tetR} \\ x_{lacI} \\ x_{cI} \\ x_{eyfp} \\ r_{tetR}(x_{cI}) \\ r_{lacI}(x_{tetR}) \\ r_{cI}(x_{lacI}) \\ r_{eyfp}(x_{cI}) \end{bmatrix}
+ \begin{bmatrix} v_1 \\ v_2 \\ v_3 \\ v_4 \end{bmatrix} \right) dt \\[2em]
& + \begin{bmatrix}
\Delta\kappa_{tetR,0} & -\Delta\gamma_{tetR} & \Delta\kappa_{tetR} \\
0 & 0 & 0 \\
0 & 0 & 0 \\
0 & 0 & 0
\end{bmatrix}
\begin{bmatrix} 1 \\ x_{tetR} \\ r_{tetR}(x_{cI}) \end{bmatrix} dw_1
+ \begin{bmatrix}
0 & 0 & 0 \\
\Delta\kappa_{lacI,0} & -\Delta\gamma_{lacI,0} & \Delta\kappa_{lacI} \\
0 & 0 & 0 \\
0 & 0 & 0
\end{bmatrix}
\begin{bmatrix} 1 \\ x_{lacI} \\ r_{lacI}(x_{tetR}) \end{bmatrix} dw_2 \\[2em]
& + \begin{bmatrix}
0 & 0 & 0 \\
0 & 0 & 0 \\
\Delta\kappa_{cI,0} & -\Delta\gamma_{cI} & \Delta\kappa_{cI} \\
0 & 0 & 0
\end{bmatrix}
\begin{bmatrix} 1 \\ x_{cI} \\ r_{cI}(x_{lacI}) \end{bmatrix} dw_3
+ \begin{bmatrix}
0 & 0 & 0 \\
0 & 0 & 0 \\
0 & 0 & 0 \\
\Delta\kappa_{eyfp,0} & -\Delta\gamma_{eyfp} & \Delta\kappa_{eyfp}
\end{bmatrix}
\begin{bmatrix} 1 \\ x_{eyfp} \\ r_{eyfp}(x_{cI}) \end{bmatrix} dw_4
\end{aligned}
\tag{3.21}
$$

Our robust synthetic gene network requires designing these parameters κ_i and γ_i within $N \in [N_1, N_2]$ to meet the four specifications, i.e., we want to design four kinetic parameters κ_{tetR}, κ_{lacI}, κ_{cI}, and κ_{eyfp} and four decay rates γ_{tetR}, γ_{lacI}, γ_{cI}, and γ_{eyfp} to satisfy the following four design specifications.

i) Suppose the biological allowable ranges of kinetic parameters and decay rates to be designed are given by Batt et al. (Batt et al. 2007)

$$
\begin{aligned}
\kappa_{tetR} &\in [50, 5000], & \kappa_{lacI} &\in [70, 7000], \\
\kappa_{cI} &\in [75, 8000], & \kappa_{eyfp} &\in [30, 30000], \\
\gamma_{tetR} &\in [0.2, 4.8], & \gamma_{lacI} &\in [0.02, 0.14], \\
\gamma_{cI} &\in [0.6, 0.8], & \gamma_{eyfp} &\in [0.1, 1].
\end{aligned}
\tag{3.22}
$$

where the allowable ranges of kinetic parameters and decay rates depend on the possibility of implementation and the biological property such as the desired steady state x_d.

ii) The standard deviations of parameter fluctuations to be tolerated are given as

$$
\begin{aligned}
\left[\Delta\kappa_{tetR,0}, \Delta\gamma_{tetR}, \Delta\kappa_{tetR} \right] &= [30, 0.3, 50], \\
\left[\Delta\kappa_{lacI,0}, \Delta\gamma_{lacI}, \Delta\kappa_{lacI} \right] &= [50, 0.3, 200], \\
\left[\Delta\kappa_{cI,0}, \Delta\gamma_{cI}, \Delta\kappa_{cI} \right] &= [30, 0.3, 50], \\
\left[\Delta\kappa_{eyfp,0}, \Delta\gamma_{eyfp}, \Delta\kappa_{eyfp} \right] &= [50, 0.3, 200].
\end{aligned}
\tag{3.23}
$$

iii) The desired steady state x_d is given by Batt et al. (Batt et al. 2007)

$$x_d = \begin{bmatrix} x_{tetR,d} \\ x_{lacI,d} \\ x_{cI,d} \\ x_{eyfp,d} \end{bmatrix} = \begin{bmatrix} 1000 \\ 30000 \\ 300 \\ 30000 \end{bmatrix} \tag{3.24}$$

iv) The prescribed attenuation level of external disturbance is specified by $\rho = 0.3$.

Based on the design procedure, we first shift the desired steady state x_d of the synthetic gene system to the origin, then perform the global linearization to obtain F_i and G_{ij} for $i = 1, ..., 4$, $j = 1, ..., 3$ (see Appendix 3.4), and finally solve LMIs for fine tuning parameters. The allowable range $[N_1, N_2]$ has been obtained by the parameter-range specification in (i). In order to simplify the selection process of N, we get $N_0 = \dfrac{1}{2}(N_1 + N_2)$ and $\tilde{N} = \dfrac{1}{2}(N_2 - N_1)$ as

$$N_0 = \begin{bmatrix} 150 & -2.5 & 0 & 0 & 0 & 2525 & 0 & 0 & 0 \\ 587 & 0 & -0.08 & 0 & 0 & 0 & 3535 & 0 & 0 \\ 210 & 0 & 0 & -0.7 & 0 & 0 & 0 & 4037.5 & 0 \\ 3487 & 0 & 0 & 0 & -0.55 & 0 & 0 & 0 & 15015 \end{bmatrix},$$

$$\tilde{N} = \begin{bmatrix} 0 & -2.3 & 0 & 0 & 0 & 2475 & 0 & 0 & 0 \\ 0 & 0 & -0.06 & 0 & 0 & 0 & 3415 & 0 & 0 \\ 0 & 0 & 0 & -0.1 & 0 & 0 & 0 & 3962.5 & 0 \\ 0 & 0 & 0 & 0 & -0.45 & 0 & 0 & 0 & 14985 \end{bmatrix}.$$

By solving LMIs for fine tuning ΔN from the allowable range $(-\tilde{N}, \tilde{N})$, we find a positive definite matrix P of LMIs in (3.20) if the allowable range is distributed over $[-\Delta N, \Delta N]$ with

$$\Delta N = \begin{bmatrix} 0 & -1.53 & 0 & 0 & 0 & 1634 & 0 & 0 & 0 \\ 0 & 0 & -0.04 & 0 & 0 & 0 & 2287 & 0 & 0 \\ 0 & 0 & 0 & -0.05 & 0 & 0 & 0 & 2615 & 0 \\ 0 & 0 & 0 & 0 & -0.28 & 0 & 0 & 0 & 9891 \end{bmatrix}$$

i.e., if the design kinetic parameters κ_i and decay rates γ_i of the synthetic gene network are specified within the following ranges:

$$\kappa_{tetR} \in [891, 4159], \qquad \kappa_{lacI} \in [1248, 5822]$$

$$\kappa_{cI} \in [1422, 6652], \qquad \kappa_{eyfp} \in [5124, 24906]$$

$$\gamma_{tetR} \in [0.97, 4.03], \qquad \gamma_{lacI} \in [0.04, 0.08] \tag{3.25}$$

$$\gamma_{cI} \in [0.65, 0.75], \qquad \gamma_{eyfp} \in [0.27, 0.83]$$

then the four design specifications (i)–(iv) are satisfied.

In order to confirm the performance of the proposed robust synthetic gene network, we design the synthetic gene network with the set of kinetic parameters κ_i and decay rates γ_i in the ranges given in (3.25) to see if they can achieve the desired steady state in spite of initial conditions, parameter fluctuations and extrinsic disturbances. Let us choose the following design parameters from the ranges given in (3.25).

$$\left(\kappa_{tetR}, \kappa_{lacI}, \kappa_{cI}, \kappa_{eyfp}\right) = (2000, 2000, 2000, 15000)$$

$$\left(\gamma_{tetR}, \gamma_{lacI}, \gamma_{cI}, \gamma_{eyfp}\right) = (1.98, 0.05, 0.7, 0.57) \tag{3.26}$$

The desired steady states can be achieved under intrinsic parameter fluctuations and extrinsic disturbances by the proposed robust synthetic gene network design method. From the simulation in Figure 3.1a with $v(t) = [10n_1, 1000n_2, 10n_3, 1000n_4]$, where n_i, $i = 1, \ldots, 4$ are independent Gaussian white noises with unit variance, the disturbance attenuation level of external disturbance, which is prescribed by $\rho = 0.3$, is estimated as

$$\frac{\left(E \int_0^{1000} \tilde{x}^T Q \tilde{x} \, dt\right)^{1/2}}{\left(E \int_0^{1000} v^T v \, dt\right)^{1/2}} = 0.2715 < 0.3$$

Clearly, the prescribed disturbance attenuation (filtering ability) is achieved by the proposed method.

In contrast to the above design case, we also design the synthetic gene network with parameters outside the ranges in (3.25), for example, with kinetic parameters $\kappa_i = (150, 100, 500, 1500)$ and decay rates $\gamma_i = (0.5, 0.05, 0.5, 0.2)$, which are outside the specified regions in (3.25). The simulation is shown in Figure 3.1b. Obviously, the time response of the synthetic network suffers more external disturbances and cannot achieve

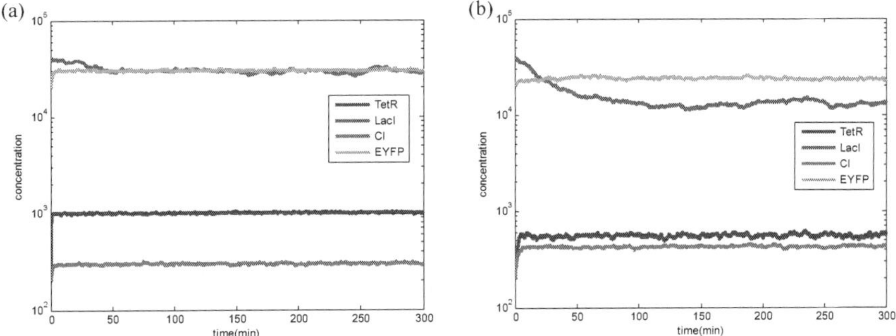

Figure 3.1. Simulation results for synthetic transcriptional cascade network. (a) With the design parameters in the specified parameter range given in (3.25). (b) With the design parameters outside the specified parameter range.

Color image of this figure appears in the color plate section at the end of the book.

the desired steady states. In this design case, the disturbance attenuation level of external disturbance is estimated as:

$$\frac{\left(E\int_{0}^{1000}\tilde{x}^{T}Q\tilde{x}dt\right)^{1/2}}{\left(E\int_{0}^{1000}v^{T}vdt\right)^{1/2}} = 2.8093 > 0.3$$

Clearly, the design specification of filtering ability is violated significantly.

From the simulation results, it can be seen that the designed synthetic gene network using the proposed method has robust stability to tolerate intrinsic parameter fluctuations and enough filtering ability to attenuate the external disturbances, thereby achieving the desired steady states. If the designed gene network has enough robust stability and filtering ability, then it could work properly under intrinsic fluctuations and extrinsic molecular noises on the host cell. Furthermore, the design ranges of kinetic parameters and decay rates can be easily solved by fine tuning ΔN in the design procedure using the LMI Toolbox in Matlab.

Appendix 3.1: Proof of proposition 3.1

Let us choose a Lyapunov function $V(\tilde{x}) > 0$ with

$$E\int_{0}^{\infty}\tilde{x}^{T}Q\tilde{x}dt = E\left\{V\left(\tilde{x}(0)\right) - V\left(\tilde{x}(\infty)\right) + \int_{0}^{\infty}\left[\tilde{x}^{T}Q\tilde{x} + \frac{dV(\tilde{x})}{dt}\right]dt\right\} \qquad (3.27)$$

By the Ito's formula (Chen and Hsu 1995) and $E\left(\dfrac{d}{dt}W_i\right)=E\left(n_i\right)=0$, we get

$$E\frac{dV(\tilde{x})}{dt}=E\left\{\left(\frac{\partial V(\tilde{x})}{\partial\tilde{x}}\right)^{T}\left[Nf(\tilde{x}+x_d)+v\right]\right.$$
$$\left.+\frac{1}{2}\sum_{i=1}^{m}g_i^{T}(\tilde{x}+x_d)M_i^{T}\frac{\partial^2 V(\tilde{x})}{\partial\tilde{x}^2}M_ig_i(\tilde{x}+x_d)\right\}$$

(3.28)

Substituting (3.28) into (3.27), we get

$$E\int_0^{\infty}\tilde{x}^{T}Q\tilde{x}dt=E\left\{V(\tilde{x}(0))-V(\tilde{x}(\infty))+\int_0^{\infty}\left[\tilde{x}^{T}Q\tilde{x}\right.\right.$$
$$+\left(\frac{\partial V(\tilde{x})}{\partial\tilde{x}}\right)^{T}Nf(\tilde{x}+x_d)+\left(\frac{\partial V(\tilde{x})}{\partial\tilde{x}}\right)^{T}v$$
$$\left.\left.+\frac{1}{2}\sum_{i=1}^{m}g_i^{T}(\tilde{x}+x_d)M_i^{T}\frac{\partial^2 V(\tilde{x})}{\partial\tilde{x}^2}M_ig_i(\tilde{x}+x_d)\right]dt\right\}$$

(3.29)

By the inequality (3.5), we have

$$E\int_0^{\infty}\tilde{x}^{T}Q\tilde{x}dt<E\left\{V(\tilde{x}(0))-V(\tilde{x}(\infty))\right.$$
$$\left.+\int_0^{\infty}\left[\left(\frac{\partial V(\tilde{x})}{\partial\tilde{x}}\right)^{T}v-\frac{1}{4\rho^2}\left(\frac{\partial V(\tilde{x})}{\partial\tilde{x}}\right)^{T}\frac{\partial V(\tilde{x})}{\partial\tilde{x}}\right]dt\right\}$$
$$\leq E\left\{V(\tilde{x}(0))+\int_0^{\infty}\left[\left(\frac{\partial V(\tilde{x})}{\partial\tilde{x}}\right)^{T}v-\frac{1}{4\rho^2}\left(\frac{\partial V(\tilde{x})}{\partial\tilde{x}}\right)^{T}\frac{\partial V(\tilde{x})}{\partial\tilde{x}}\right]dt\right\}$$
$$\leq E\left\{V(\tilde{x}(0))+\int_0^{\infty}\left[\rho^2 v^{T}v\right.\right.$$

(3.30)

$$\left.\left.-\left(\rho v-\frac{1}{2\rho}\frac{\partial V(\tilde{x})}{\partial\tilde{x}}\right)^{T}\left(\rho v-\frac{1}{2\rho}\frac{\partial V(\tilde{x})}{\partial\tilde{x}}\right)\right]dt\right\}$$
$$\leq E\left\{V(\tilde{x}(0))+\int_0^{\infty}\rho^2 v^{T}vdt\right\}$$

If $V(\tilde{x}(0)) = 0$, then (3.30) will be reduced to (3.2).

If the synthetic gene network is free of the external disturbances, i.e., $v = 0$, then we have

$$E \int_0^\infty \tilde{x}^T Q \tilde{x}\, dt < V(\tilde{x}(0)) \tag{3.31}$$

For some positive constant $V(\tilde{x}(0))$, it means $\tilde{x} \to 0$ in probability as $t \to \infty$.

Appendix 3.2: Proof of proposition 3.2

Let us choose a Lyapunov function $V(\tilde{x}) > 0$ for the stochastic gene network in (3.7). By the Ito formula (Chen and Hsu 1995), we get

$$
\begin{aligned}
E \frac{dV(\tilde{x})}{dt} = E \Bigg\{ & \left(\frac{\partial V(\tilde{x})}{\partial \tilde{x}} \right)^T \left[Nf(\tilde{x} + x_d) + v \right] \\
& + \frac{1}{2} \sum_{i=1}^m g_i^T (\tilde{x} + x_d) M_i^T \frac{\partial^2 V(\tilde{x})}{\partial \tilde{x}^2} M_i g_i (\tilde{x} + x_d) \\
& + \left(\frac{\partial V(\tilde{x})}{\partial \tilde{x}} \right)^T N \Delta f(\tilde{x} + x_d) \\
& + \frac{1}{2} \sum_{i=1}^m \Delta g_i^T (\tilde{x} + x_d) M_i^T \frac{\partial^2 V(\tilde{x})}{\partial \tilde{x}^2} M_i \Delta g_i (\tilde{x} + x_d) \Bigg\}
\end{aligned} \tag{3.32}
$$

By the fact that

$$
\begin{aligned}
E \left[\left(\frac{\partial V(\tilde{x})}{\partial \tilde{x}} \right)^T N \Delta f(\tilde{x} + x_d) \right] \le{} & \frac{1}{4} \left(\frac{\partial V(\tilde{x})}{\partial \tilde{x}} \right)^T NN^T \left(\frac{\partial V(\tilde{x})}{\partial \tilde{x}} \right) \\
& + E \left[\Delta f^T (\tilde{x} + x_d) \Delta f(\tilde{x} + x_d) \right] \\
\le{} & \frac{1}{4} \left(\frac{\partial V(\tilde{x})}{\partial \tilde{x}} \right)^T NN^T \left(\frac{\partial V(\tilde{x})}{\partial \tilde{x}} \right) + \alpha^2 E(\tilde{x} + x_d)^T (\tilde{x} + x_d)
\end{aligned} \tag{3.33}
$$

and $E\left[\Delta g_i^T (\tilde{x} + x_d) \Delta g_i (\tilde{x} + x_d) \right] \le \beta_i^2 E(\tilde{x} + x_d)^T (\tilde{x} + x_d)$, we get the following inequality after some rearrangements

$$
E\frac{dV(\tilde{x})}{dt} \leq E\left\{\left(\frac{\partial V(\tilde{x})}{\partial \tilde{x}}\right)^T \left[Nf(\tilde{x}+x_d)+v\right]\right.
$$

$$
+\frac{1}{2}\sum_{i=1}^{m} g_i^T(\tilde{x}+x_d) M_i^T \frac{\partial^2 V(\tilde{x})}{\partial \tilde{x}^2} M_i g_i(\tilde{x}+x_d)
$$

$$
+\frac{1}{4}\left(\frac{\partial V(\tilde{x})}{\partial \tilde{x}}\right)^T NN^T\left(\frac{\partial V(\tilde{x})}{\partial \tilde{x}}\right)+\alpha^2(\tilde{x}+x_d)^T(\tilde{x}+x_d)
$$

$$
\left.+\frac{1}{2}\sum_{i=1}^{m}\beta_i^2(\tilde{x}+x_d)^T M_i^T \frac{\partial^2 V(\tilde{x})}{\partial \tilde{x}^2} M_i(\tilde{x}+x_d)\right\}
\tag{3.34}
$$

Then, following the same procedure as the proof of Proposition 3.1 in Appendix 3.1, we can get the results of Proposition 3.2, i.e., we can get the same result as Proposition 3.1 except three extra terms due to the need to tolerate functional perturbations in (3.34).

Appendix 3.3: Proof of proposition 3.3

Now, we will derive the sufficient condition to ensure that the interpolated linear synthetic gene network in (3.12) can attenuate the external disturbance below a prescribed attenuation level ρ in (3.2) or (3.3). By choosing a positive Lyapunov function as $V(\tilde{x})=\tilde{x}^T P\tilde{x}>0$, we have

$$
E\int_0^{\infty}\tilde{x}^T Q\tilde{x}\,dt = E\left\{\tilde{x}^T(0)P\tilde{x}(0)-\tilde{x}^T(\infty)P\tilde{x}(\infty)\right.
$$

$$
\left.+\int_0^{\infty}\left[\tilde{x}^T Q\tilde{x}+\frac{d}{dt}\tilde{x}^T P\tilde{x}\right]dt\right\}
\tag{3.35}
$$

By Ito's formula and $E\left(\dfrac{dW_i}{dt}\right)=0$, we have

$$E\int_0^\infty \tilde{x}^T Q\tilde{x}\,dt = E\left\{\tilde{x}^T(0)P\tilde{x}(0) - \tilde{x}^T(\infty)P\tilde{x}(\infty) + \int_0^\infty \left[\tilde{x}^T Q\tilde{x}\right.\right.$$

$$+\sum_{j=1}^M \alpha_j(\tilde{x})\left(NF_j\tilde{x}+v\right)^T P\tilde{x} + \sum_{j=1}^M \alpha_j(\tilde{x})\tilde{x}^T P\left(NF_j\tilde{x}+v\right)$$

$$\left.\left.+\sum_{j=1}^M \alpha_j(\tilde{x})\sum_{i=1}^m \tilde{x}^T G_{ij}^T M_i^T PM_i G_{ij}\tilde{x}\right]dt\right\}$$

$$\le E\left\{\tilde{x}^T(0)P\tilde{x}(0) + \int_0^\infty \sum_{j=1}^M \alpha_j(\tilde{x})\tilde{x}^T\left(F_j^T N^T P + PNF_j + Q\right.\right.$$

$$\left.+\sum_{i=1}^m G_{ij}^T M_i^T PM_i G_{ij} + \frac{1}{\rho^2}PP\right)\tilde{x}\,dt + \sum_{j=1}^M \alpha_j(\tilde{x})\left(\tilde{x}^T Pv + v^T P\tilde{x}\right.$$

$$\left.\left.-\frac{1}{\rho^2}\tilde{x}^T PP\tilde{x} - \rho^2 v^T v\right) + \rho^2 v^T v\,dt\right\}$$

$$(3.36)$$

By the inequality in (3.13), we have

$$E\int_0^\infty \tilde{x}^T Q\tilde{x}\,dt \le E\left\{\tilde{x}^T(0)P\tilde{x}(0) + \int_0^\infty \rho^2 v^T v\,dt\right.$$

$$\left.-\int_0^\infty\left[\sum_{j=1}^M \alpha_j(\tilde{x})\left(\frac{1}{\rho}\tilde{x}^T P - \rho v\right)^T\left(\frac{1}{\rho}\tilde{x}^T P - \rho v\right)\right]dt\right\}$$

$$(3.37)$$

$$\le E\left\{\tilde{x}^T(0)P\tilde{x}(0) + \int_0^\infty \rho^2 v^T v\,dt\right\}$$

Then, the remainder of the proof is similar to the procedure in Appendix 3.1.

Appendix 3.4

The global linearization technique can be employed to transform the nonlinear stochastic gene network into an interpolation of a set of globally linearized gene networks. In this design example, the global linearizations are bound by a polytope consisting of 3 vertices, shown as follows

$$d\tilde{x} = \left(NF_j\tilde{x}+v\right)dt + \sum_{i=1}^m M_i G_{ij}\tilde{x}\,dW_i, j = 1, 2, 3$$

where

$$F_1 = \begin{bmatrix} 0 & 0 & 0 & 0 \\ 1 & 0 & 0 & 0 \\ 0 & 1 & 0 & 0 \\ 0 & 0 & 1 & 0 \\ 0 & 0 & 0 & 1 \\ 0 & 0 & -3.6982\times10^{-4} & 0 \\ -3.6982\times10^{-4} & 0 & 0 & 0 \\ 0 & -3.1211\times10^{-8} & 0 & 0 \\ 0 & 0 & -3.6982\times10^{-4} & 0 \end{bmatrix} \quad F_2 = \begin{bmatrix} 0 & 0 & 0 & 0 \\ 1 & 0 & 0 & 0 \\ 0 & 1 & 0 & 0 \\ 0 & 0 & 1 & 0 \\ 0 & 0 & 0 & 1 \\ 0 & 0 & -5.0491\times10^{-4} & 0 \\ -4.9998\times10^{-4} & 0 & 0 & 0 \\ 0 & -7.2481\times10^{-8} & 0 & 0 \\ 0 & 0 & -5.0491\times10^{-4} & 0 \end{bmatrix}$$

$$F_3 = \begin{bmatrix} 0 & 0 & 0 & 0 \\ 1 & 0 & 0 & 0 \\ 0 & 1 & 0 & 0 \\ 0 & 0 & 1 & 0 \\ 0 & 0 & 0 & 1 \\ 0 & 0 & -5.0501\times10^{-4} & 0 \\ -5\times10^{-4} & 0 & 0 & 0 \\ 0 & -7.3881\times10^{-8} & 0 & 0 \\ 0 & 0 & -5.0501\times10^{-4} & 0 \end{bmatrix}$$

$$G_{11} = \begin{bmatrix} 0 & 0 & 0 & 0 \\ 1 & 0 & 0 & 0 \\ 0 & 0 & -3.6982\times10^{-4} & 0 \end{bmatrix}, \quad G_{21} = \begin{bmatrix} 0 & 0 & 0 & 0 \\ 0 & 0 & 1 & 0 \\ -3.6982\times10^{-4} & 0 & 0 & 0 \end{bmatrix}$$

$$G_{31} = \begin{bmatrix} 0 & 0 & 0 & 0 \\ 0 & 0 & 1 & 0 \\ 0 & -3.1211\times10^{-8} & 0 & 0 \end{bmatrix}, \quad G_{41} = \begin{bmatrix} 0 & 0 & 0 & 0 \\ 0 & 0 & 0 & 1 \\ 0 & 0 & -3.6982\times10^{-4} & 0 \end{bmatrix}$$

$$G_{12} = \begin{bmatrix} 0 & 0 & 0 & 0 \\ 1 & 0 & 0 & 0 \\ 0 & 0 & -5.0491\times10^{-4} & 0 \end{bmatrix}, \quad G_{22} = \begin{bmatrix} 0 & 0 & 0 & 0 \\ 0 & 0 & 1 & 0 \\ -4.9998\times10^{-4} & 0 & 0 & 0 \end{bmatrix}$$

$$G_{32} = \begin{bmatrix} 0 & 0 & 0 & 0 \\ 0 & 0 & 1 & 0 \\ 0 & -7.2481\times10^{-8} & 0 & 0 \end{bmatrix}, \quad G_{42} = \begin{bmatrix} 0 & 0 & 0 & 0 \\ 0 & 0 & 0 & 1 \\ 0 & 0 & -5.0491\times10^{-4} & 0 \end{bmatrix}$$

$$G_{13} = \begin{bmatrix} 0 & 0 & 0 & 0 \\ 1 & 0 & 0 & 0 \\ 0 & 0 & -5.0501\times10^{-4} & 0 \end{bmatrix}, \quad G_{23} = \begin{bmatrix} 0 & 0 & 0 & 0 \\ 0 & 0 & 1 & 0 \\ -5\times10^{-4} & 0 & 0 & 0 \end{bmatrix}$$

$$G_{33} = \begin{bmatrix} 0 & 0 & 0 & 0 \\ 0 & 0 & 1 & 0 \\ 0 & -7.3881\times10^{-8} & 0 & 0 \end{bmatrix}, \quad G_{34} = \begin{bmatrix} 0 & 0 & 0 & 0 \\ 0 & 0 & 0 & 1 \\ 0 & 0 & -5.0501\times10^{-4} & 0 \end{bmatrix}$$

References

Alon, U. 2007. An Introduction to Systems Biology: Design Principles of Biological Circuits. Chapman & Hall/CRC, London.

Batt, G., Yordanov, B., Weiss, R. and Belta, C. 2007. Robustness analysis and tuning of synthetic gene networks. Bioinformatics 23: 2415.

Boyd, S., El Ghaoui, L., Feron, E. and Balakrishnan, V. 1994. Linear Matrix Inequalities in System and Control Theory. Society for Industrial Mathematics, Philadelphia.

Chen, B.S. and Zhang, W. 2004. Stochastic H-2/H-infinity control with state-dependent noise. IEEE Transactions on Automatic Control 49: 45–57.

Chen, B.S., Chang, Y.T. and Wang, Y.C. 2008. Robust H infinity-stabilization design in gene networks under stochastic molecular noises: fuzzy-interpolation approach. IEEE Trans Syst Man Cybern B Cybern 38: 25–42.

Chen, B.S. and Wu, C.H. 2009. A systematic design method for robust synthetic biology to satisfy design specifications. BMC Syst Biol 3: 66.

Chen, G. and Hsu, S.-H. 1995. Linear Stochastic Control Systems. Boca Raton, FL: CRC Press.

Drulhe, S., Ferrari-Trecate, G. and de Jong, H. 2008. The switching threshold reconstruction problem for piecewise affine models of genetic regulatory networks. IEEE Automatic Control 53: 153–165.

Gahinet, P., Nemirovski, A., Laub, A.J. and Chilali, M. 1995. LMI Control Toolbox User's Guide. The MathWorks, Inc., Natick, MA.

Slotine, J.-J.E. and Li, W. 1991. Applied nonlinear control. Prentice Hall Englewood Cliffs, NJ.

4

Robust Synthetic Biology Designs based on Network Evolutionary Methods

4.1 Robust Synthetic Biology Design using GA and EA

In the above robust synthetic gene network designs, we need to solve a set of LMIs in (3.20). In this chapter, a simple but sufficient robust synthetic gene network design method via evolutionary algorithms is introduced. To mimic the natural selection in evolution in order to select adequate design parameters for obtaining a robust synthetic gene network with desired behaviors under intrinsic parameter fluctuations and extrinsic disturbances on the host cell, the proposed network evolutionary design method based on genetic algorithm (GA) or evolutionary algorithm (EA) can search for design parameters to achieve the fitness maximization which is equivalent to the optimal tracking of desired behavior under the effect of intrinsic and extrinsic noises on the host cell.

Suppose a general synthetic gene network in (2.6) with intrinsic parameter fluctuations and extrinsic noises *in vivo* can be represented as a set of nonlinear stochastic equation in the form

$$\dot{x} = f(x,k,u) + \sum_{i=1}^{m} h_i(x,\Delta k)n_i + v$$

$$y = cx$$

(4.1)

where f is a nonlinear nominal interaction vector function concerning the state vector $x = [x_1,...,x_n]^T$ for concentrations of n reactant species, kinetic constants k and input signals u. n_i are the independent intracellular random fluctuation sources. h_i are fluctuation functions due to random fluctuation source n_i. $v = [v_1,...,v_n]^T$ represents the vector of external disturbances. y stands for the output vector.

In real biological system, gene network in (4.1) could evolve adaptively with kinetic parameters in k by natural selection through mutation and genetic variation so that $y(t)$ can robustly achieve some desired behavior $y_d(t)$ in Figure 4.1 in spite of intracellular molecular noise and external disturbance. Here, we will mimic the evolutionary biological system to adapt the kinetic parameters of synthetic gene network in (4.1) through a network evolutionary method via GA or EA under a fitness function so that the synthetic gene network can achieve a desired behavior in spite of intracellular noise and external disturbance specified forehand to be tolerated *in vivo*.

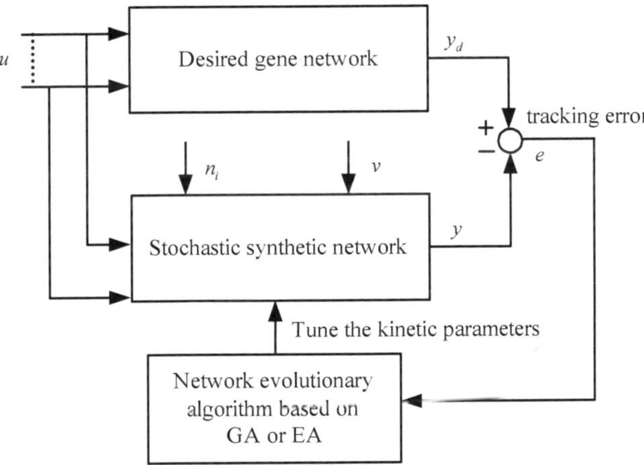

Figure 4.1. The block diagram of robust synthetic biology designs based on network evolutionary methods.

Based on the analysis above, given the design specifications, our design objective is to tune the design parameters $k_i \in \left[k_i^L, k_i^U \right]$, $i = 1, 2, 3,...,l$ to achieve optimal tracking under intrinsic parametric fluctuations and extrinsic noises as shown in Figure 4.1. Suppose the tracking error is defined as $e = y_d - y$, where y_d denotes the output of the desired synthetic gene network (for example, logic network or oscillator network). Then, our design purpose is to tune design parameter k_i by GA or EA algorithm so that the stochastic synthetic gene network can achieve the following optimal tracking (Chen and Chen 2010, Chen et al. 2011)

$$\min_{\substack{k_i \in [k_i^L, k_i^U] \\ i=1,...,l}} \mathrm{E} \int_0^{T_p} e^T(t)e(t)dt \tag{4.2}$$

where E denotes the expectation, T_p denotes the present time and $e^T(t)$ denotes the transpose of $e(t)$. If the above mean square tracking error can be minimized, the robust synthetic gene network can track the desired network more reliably in the host cells. Here, to mimic the parametric tuning of biological network to achieve a desired function via natural selection in evolution, an artificial GA or EA is employed to tune the design parameter k_i to achieve the optimal tracking in (4.2) but with a faster speed than natural selection (Chen and Chen 2010, Chen et al. 2011). GAs and EAs are results of an effort to model adaptation phenomena in natural and artificial systems. These GAs and EAs will be modified to tune the kinetic parameters of nonlinear stochastic synthetic gene network in (4.1), that is, the so-called network evolutionary algorithm, to fast evolve to a desired output behavior via a fitness function. In the nonlinear stochastic system of synthetic gene network in (4.1), the state vector x is considered as phenotype and the kinetic parameter vector $k = [k_1, ..., k_l]^T$ is considered as genotype. In the network evolutionary algorithm, the kinetic parameter vector k is called chromosome. Let us denote the mean square error in (4.2) as

$$J(k) = \mathrm{E} \int_0^{T_p} e^T(t)e(t)dt \tag{4.3}$$

where the chromosome $k \in \left[k^L, k^U \right]$, the feasible parameter space or feasible genotype space. Define the fitness function $F(k)$ as

$$F(k) = \frac{1}{J(k)} \tag{4.4}$$

that is, a small mean square error means a large fitness and vice versa. If we adapt a parameter vector (chromosome) $k \in \left[k^L, k^U \right]$ by GA or EA to minimize $J(k)$ in (4.3) or (4.4), then we achieve the maximization of fitness function in (4.4) for synthetic gene network (4.1) to meet the natural selection in evolution. Therefore, the robust synthetic gene network in (4.1) with a desired output behavior $y_d(t)$ is equivalent to solve the following fitness maximization problem by the proposed network evolutionary method (Chen and Chen 2010, Chen et al. 2011).

$$F(k^*) = \max_{k \in [k^L, k^U]} F(k) \tag{4.5}$$

The network evolutionary algorithm based on GAs or EAs is employed to solve the above fitness maximization problem via genetic operators such as selection, crossover, and mutation to mimic the natural selection in the evolutionary process to tune the kinetic parameter vector k of synthetic gene network to solve the optimization problem in (4.5) to achieve robust optimal tracking of the desired behavior. A simple network evolutionary algorithm is proposed as follows:

1. Initialization. Initialize a population of candidate solutions to the problem, that is, randomly generate a population of candidate chromosomes. In the real coding representation, each chromosome with the same length as the vector of decision variables is encoded as a vector of floating-point numbers. The vector $k = (k_1,...,k_l)$ is as a chromosome to represent a solution of optimization problem for the desired behavior tracking of nonlinear stochastic synthetic gene network. Initialization procedure produces M chromosomes $k^1,...,k^i,...,k^M$, where M denotes the population size.

2. Fitness. Fitness is a measure to evaluate the suitability of chromosome. By the principle of survival of the fittest, a chromosome with higher fitness value has a higher probability of contributing one or more offspring in the next generation. By employing network evolutionary algorithms to our fitness optimization problem, we must relate the M chromosomes with their fitness functions $F(k^1),...,F(k^M)$. In our synthetic gene network design problem, an optimal tracking design is to select a maximum fitness function $F(k^*)$ among these fitness functions $F(k^1),...,F(k^i)...,F(k^M)$.

3. Reproduction. Reproduction is a basic operator of GAs or EAs to generate more offspring to increase the possibility to search for the optimal fitness. It is operated on the basis of the survival of the fitness. In each generation, the chromosomes of the current population are reproduced or copied in the next generation according to their reproduction probability p_{r_i}, which are defined as

$$p_{r_i} = \frac{F(k^i)}{\sum_{i=1}^{M} F(k^i)}, \quad i = 1,...,M \tag{4.6}$$

where M is the population size. It is shown that the higher fitness value, the higher reproduction probability. Once the chromosomes are reproduced or copied in the next generation, the other chromosomes stay in a mating pool as shown in Figure 4.2 and await the action of the other two genetic operators.

4. Crossover. Crossover provides a mechanism for strings to mix and match their desirable qualities through a random process. Since GAs work with a population of binary strings, the crossover proceeds in three steps. First, two newly reproduced strings are selected from the mating pool produced by reproduction. Second, a position along the

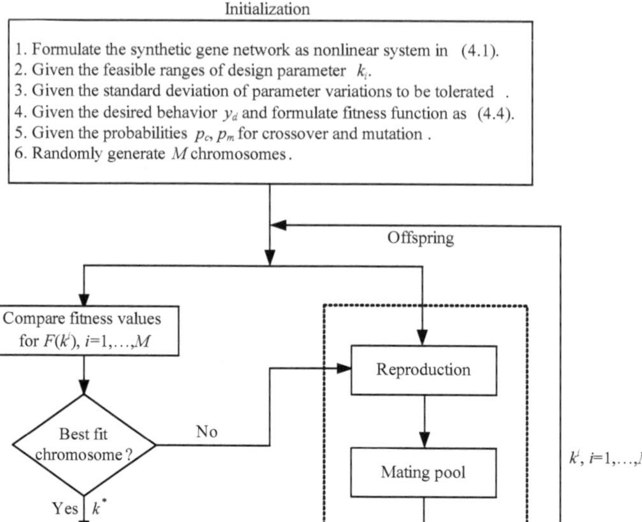

Figure 4.2. The flow chart of network evolutionary algorithm for solving the best fit k^* from the maximization problem of fitness function in (4.5).

two strings is selected uniformly at random. This is illustrated below where two binary coded strings, $(\alpha_1, \alpha_2, \alpha_3)_i$ and $(\alpha_1, \alpha_2, \alpha_3)_j$ with length $l = 12$ are shown aligned for crossover.

$$\downarrow \text{Crossing site}$$

$$k^i = \left(\alpha_1, \alpha_2, \alpha_3\right)_i = 1100\!:\!00011111$$

$$k^j = \left(\alpha_1, \alpha_2, \alpha_3\right)_j = 0010\!:\!11001011$$

The third step is to exchange all characters following the crossing site. For example, the two strings $(\alpha_1, \alpha_2, \alpha_3)_i$ and $(\alpha_1, \alpha_2, \alpha_3)_j$ with a crossover at 4th bit become

$$k^i = \left(\alpha_1, \alpha_2, \alpha_3\right)_i = 1100\!:\!11001011$$

$$k^j = \left(\alpha_1, \alpha_2, \alpha_3\right)_j = 0010\!:\!00011111$$

Although crossover uses random choice, it should not be thought of as a random walk through the search space. When combined with reproduction, it is an effective means of exchanging information and combining portions of high-quality solutions. In contrast to GAs using the binary representation, EAs work on real-valued vectors for optimization algorithm, which can make the search process for a global optimizer easier. For the chromosome k^i and k^j randomly selected according to crossover probability p_c, the resulting offspring k' is (Chen et al. 2011)

$$k' = r(k^i + k^j)$$

where $r \in (0,1)$.

5. Mutation. Reproduction and crossover provide the most search power for GAs or EAs. However, the mating pool tends to become more and more homogeneous as one better solution begins to dominate after several generations and leads to premature convergence. In the situation, the third operator, mutation, is introduced into the GA or EA with appropriate probability p_m. In GA, an example is given to illustrate the mutation.

$$\downarrow \text{Mutation}$$

$$k^i = (\alpha_1, \alpha_2, \alpha_3)_i = 10100101[0]101$$

$$k^i = (\alpha_1, \alpha_2, \alpha_3)_i = 10100101[1]101$$

In the case of binary code, the mutation operator simply flips the state of a bit from 0 to 1 at the 9th code or vice versa. In the EA, for a given chromosome $k = (k_1, k_2, ..., k_n, ..., k_l)$, if the element k_n is randomly selected for mutation, the resulting offspring is $k' = (k_1, k_2, ..., k'_n, ..., k_l)$. The new gene k'_n is

$$k'_n = k_n + \sigma_n m_n$$

where σ_n is standard derivation and m_n is a random variable with standard normal distribution function. Note that mutation should be used sparingly because it is inherently a random search operator. EAs and GAs could become more similar to random search if the mutation probability is high.

Since the proposed network evolutionary method not only achieve the best fitness for optimal desired behavior tracking in (4.5) but also robustly tolerate random kinetic parameter fluctuation and external disturbance simultaneously, it will play an important role for synthetic gene network from network evolutionary perspective. The network evolutionary approach of how to select a parameter vector (or chromosome) k to solve the fitness maximization problem in (4.5) for the optimal behavior tracking is summarized as follows (see Figure 4.2) (Chen and Chen 2010, Chen et al. 2011).

Step 1. Given the design specifications

Step 2. Model the synthetic gene network as the nonlinear stochastic system in (4.1)

Step 3. Specify the crossover probabilities p_c, mutation probability p_m and the population size M for GA or EA

Step 4. Generate randomly a population of candidate chromosomes

Step 5. Evaluate the fitness $F(k^i)$ for each candidate solution (chromosome) k^i in the population to find the best fit k^* to achieve the best fitness $F(k^*)$

Step 6. If the search goal is achieved, or an allowable generation is attained, then stop. Otherwise, continue

Step 7. Replace the current population with a new population by applying selection, crossover, and mutation operations on the current population. Go to Step 5.

In order to illustrate the design procedure of the proposed robust synthetic gene oscillators by GA or EA search method, the following two examples with numerical simulations are given to describe the design procedure.

4.2 An *in silico* Design Example: GA Approach

Consider the synthetic genetic oscillator shown in Figure 4.3.

The first repressor protein, LacI from *E. coli*, inhibits the transcription of the second repressor gene, *tetR* from the tetracycline-resistance transposon

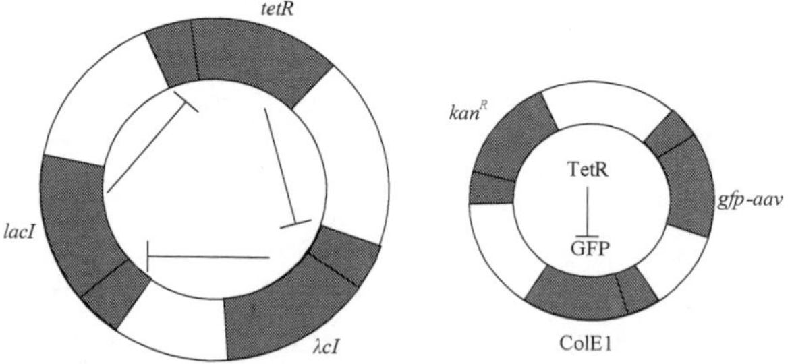

Figure 4.3. Construction of the repressilator network in the host cell, *E. coli*. The repressilator is a cyclic negative-feedback loop composed of three repressor genes (the red regions) *tetR*, *λcI*, *lacI* and their corresponding promoters (the gray regions) in plasmid. The compatible reporter plasmid expresses an intermediate-stability GFP variant (gfp-aav) on the ring.

Color image of this figure appears in the color plate section at the end of the book.

Tn10, whose protein product in turn inhibits the expression of the third gene, *cI* from the λ phage. Finally, CI inhibits *lacI* expression, completing the cycle. The negative feedback loop in the following transcriptional regulatory model can lead to temporal oscillations in the concentration of each component for us to design the repressilator and study its robust dynamic behavior (Elowitz and Leibler 2000).

$$\frac{dm_i}{dt} = -\gamma_{m_i} m_i + \frac{\alpha_i}{1 + \left(p_j / K_i\right)^n} + \alpha_{0i}$$

$$\frac{dp_i}{dt} = \beta_i m_i - \gamma_{p_i} p_i$$

(4.7)

where m_i is the concentration of messenger RNA (mRNA) and p_i and p_j are concentrations of proteins for $i = 1,2,3$ corresponding to LacI, TetR, and CI; and $j = 3,1,2$ corresponding to CI, LacI, and TetR, respectively. Parameters γ_{mi} and γ_{pi} are the decay rates of mRNAs and proteins. α_i is the transcription rate of mRNA. α_{0i} is the effect of leakiness and is usually zero for stable state. β_i accounts for the number of translated protein molecules per mRNA molecule. K_i is the number of the *j*th proteins for a half repression of the *i*th promoter.

In this model, the network behavior depends on the transcription rate of repressor concentration, the translation rates and decay rates of protein and mRNA. Depending on the values of these parameters, the network may be stable, chaotic or leading to sustained limit-cycle oscillations. Oscillations are favored by gene regulatory networks with strong promoters containing an efficient ribosome-binding site, tight transcriptional repression (low 'leakiness'), cooperative repression characteristics, and comparable protein and mRNA decay rates. A further obstacle to the design of oscillatory biochemical networks is internal uncertainty, e.g., the thermal fluctuation and the stochastic effects due to the small number of particles involved, characterized as the fluctuations of parameters, and external disturbance on the host cell from the environment. These intrinsic parameter fluctuations and extrinsic molecular noises also may lead sustained oscillations to stable steady states or chaos. Although synthetic oscillators are much simpler than the real biological oscillations, at present these synthetic oscillators still cannot work reliably for a long time and need further tuning before application. It is still difficult to systematically design a synthetic gene oscillator with desired amplitude, frequency and phase specified before hand by the user. In practical applications, a robust synthetic gene oscillator with the desired amplitude, frequency and phase under intrinsic and extrinsic molecular noises is more useful. More efforts are still needed to achieve this kind of robust synthetic gene oscillator *in vivo* design.

Therefore, a robust synthetic oscillator network with desired amplitude, frequency and phase is more appealing for synthetic biologist. Before further discussion on the robust design of synthetic biological oscillators, a stochastic model for synthetic biological oscillator with intrinsic fluctuations and extrinsic disturbances *in vivo* is introduced as follows (Chen and Chen 2010)

$$\dot{m}_1 = -\left(\gamma_{m_1} + \Delta\gamma_{m_1}\right)m_1 + \frac{\left(\alpha_1 + \Delta\alpha_1\right)}{1 + \left(p_3/K_1\right)^n} + \alpha_{01} + v_1$$

$$\dot{p}_1 = \left(\beta_1 + \Delta\beta_1\right)m_1 - \left(\gamma_{p_1} + \Delta\gamma_{p_1}\right)p_1 + v_2$$

$$\dot{m}_2 = -\left(\gamma_{m_2} + \Delta\gamma_{m_2}\right)m_2 + \frac{\left(\alpha_2 + \Delta\alpha_2\right)}{1 + \left(p_1/K_2\right)^n} + \alpha_{02} + v_3$$

$$\dot{p}_2 = \left(\beta_2 + \Delta\beta_2\right)m_2 - \left(\gamma_{p_2} + \Delta\gamma_{p_2}\right)p_2 + v_4 \qquad (4.8)$$

$$\dot{m}_3 = -\left(\gamma_{m_3} + \Delta\gamma_{m_3}\right)m_3 + \frac{\left(\alpha_3 + \Delta\alpha_3\right)}{1 + \left(p_2/K_3\right)^n} + \alpha_{03} + v_5$$

$$\dot{p}_3 = \left(\beta_3 + \Delta\beta_3\right)m_3 - \left(\gamma_{p_3} + \Delta\gamma_{p_3}\right)p_3 + v_6$$

where $\Delta\gamma_{m_i}$, $\Delta\alpha_i$, $\Delta\beta_i$, and $\Delta\gamma_{pi}$ denote the kinetic parametric fluctuations and v_k denotes the corresponding external stochastic disturbances with variance σ_k^2, for $k = 1 \cdots 6$.

Suppose the parametric fluctuations are stochastic as follows

$$\Delta\gamma_{m_i} = \delta\gamma_{m_i} n_1$$

$$\Delta\alpha_i = \delta\alpha_i n_2 \qquad (4.9)$$

$$\Delta\beta_i = \delta\beta_i n_3$$

$$\Delta\gamma_{p_i} = \delta\gamma_{p_i} n_4$$

where $\delta\gamma_{mi}$, $\delta\alpha_i$, $\delta\beta_i$, and $\delta\gamma_{pi}$ denote the deterministic parts of parametric fluctuations, and n_1, n_2, n_3, and n_4 are independent standard white noises to denote the random fluctuation sources with unit variance.

$$\mathrm{var}\left(\Delta\gamma_{m_i}\right) = \left(\delta\gamma_{m_i}\right)^2$$

$$\mathrm{var}\left(\Delta\alpha_i\right) = \left(\delta\alpha_i\right)^2$$

$$\mathrm{var}\left(\Delta\beta_i\right) = \left(\delta\beta_i\right)^2$$

$$\mathrm{var}\left(\Delta\gamma_{p_i}\right) = \left(\delta\gamma_{p_i}\right)^2$$

i.e., $\delta\gamma_{m_i}$, $\delta\alpha_i$, $\delta\beta_i$, $\delta\gamma_{pi}$ denote the standard deviations of the corresponding stochastic parametric fluctuations of $\Delta\gamma_{mi}$, $\Delta\alpha_i$, $\Delta\beta_i$, $\Delta\gamma_{pi}$, respectively.

Substituting (4.9) into (4.8), we get the following stochastic synthetic oscillator *in vivo*

$$
\begin{bmatrix} \dot{m}_1 \\ \dot{p}_1 \\ \dot{m}_2 \\ \dot{p}_2 \\ \dot{m}_3 \\ \dot{p}_3 \end{bmatrix} =
\begin{bmatrix}
-\gamma_{m_1} m_1 + \dfrac{\alpha_1}{(1+p_3/K_1)} + \alpha_{01} \\
\beta_1 m_1 - \gamma_{p_1} p_1 \\
-\gamma_{m_2} m_2 + \dfrac{\alpha_2}{(1+p_1/K_2)} + \alpha_{02} \\
\beta_2 m_2 - \gamma_{p_2} p_2 \\
-\gamma_{m_3} m_3 + \dfrac{\alpha_3}{(1+p_2/K_3)} + \alpha_{03} \\
\beta_3 m_3 - \gamma_{p_4} p_3
\end{bmatrix}
-
\begin{bmatrix} \delta\gamma_{m_1} m_1 \\ 0 \\ \delta\gamma_{m_2} m_2 \\ 0 \\ \delta\gamma_{m_3} m_3 \\ 0 \end{bmatrix} n_1
+
\begin{bmatrix} \dfrac{\delta\alpha_1}{(1+p_3/K_1)} \\ 0 \\ \dfrac{\delta\alpha_2}{(1+p_1/K_2)} \\ 0 \\ \dfrac{\delta\alpha_3}{(1+p_2/K_3)} \\ 0 \end{bmatrix} n_2
+
\begin{bmatrix} 0 \\ \delta\beta_1 m_1 \\ 0 \\ \delta\beta_2 m_2 \\ 0 \\ \delta\beta_3 m_3 \end{bmatrix} n_3
-
\begin{bmatrix} 0 \\ \delta\gamma_{p_1} p_1 \\ 0 \\ \delta\gamma_{p_2} p_2 \\ 0 \\ \delta\gamma_{p_3} p_3 \end{bmatrix} n_4
+
\begin{bmatrix} v_1 \\ v_2 \\ v_3 \\ v_4 \\ v_5 \\ v_6 \end{bmatrix}
\tag{4.10}
$$

A more general form of stochastic system for synthetic biological oscillator (4.10) under intrinsic parameter fluctuations and external disturbances in the context of the host cell can be represented by the following nonlinear equation (Chen and Chen 2010)

$$
\dot{x} = f(x) + \sum_{i=1}^{M} g_i(x) n_i + v \tag{4.11}
$$

where $x = [m_1 \quad p_1 \quad \cdots \quad m_3 \quad p_3]^T$ denotes the state vector of the synthetic biological oscillator; $v = [v_1 \quad \cdots \quad v_6]^T$ denotes the external disturbance *in vivo*; $f(x)$ denotes the nonlinear biochemical interactions of synthetic biological oscillator; and $g_i(x)$ denotes the effect of the *i*th random fluctuation source n_i. In a real biological oscillator, the robust kinetic parameters are selected by natural selection in the evolutionary process to achieve robust oscillation under intrinsic and extrinsic molecular noises. In this section, we mimic the design rules of natural selection via genetic algorithm (GA) to select adequate kinetic parameters to achieve a robust genetic oscillator design via fast computer simulation.

Consider the synthetic gene oscillator example with the same parameters as (Elowitz and Leibler 2000), except that α_i and γ_{pi} are to be designed. Suppose the desired phases of three proteins are uniformly distributed, e.g., $\phi_1 = 2\pi/3$, $\phi_2 = 0$, and $\phi_3 = -2\pi/3$. Then we give a prescribed reference model for the repressilator as follows (Figure 4.4a)

$$
y = \begin{bmatrix} y_1 \\ y_2 \\ y_3 \end{bmatrix} =
\begin{bmatrix}
28\sin(0.0208\pi t + 2\pi/3) + 30 \\
27\sin(0.0208\pi t) + 31 \\
28\sin(0.0208\pi t - 2\pi/3) + 30.5
\end{bmatrix}
\tag{4.12}
$$

(a)

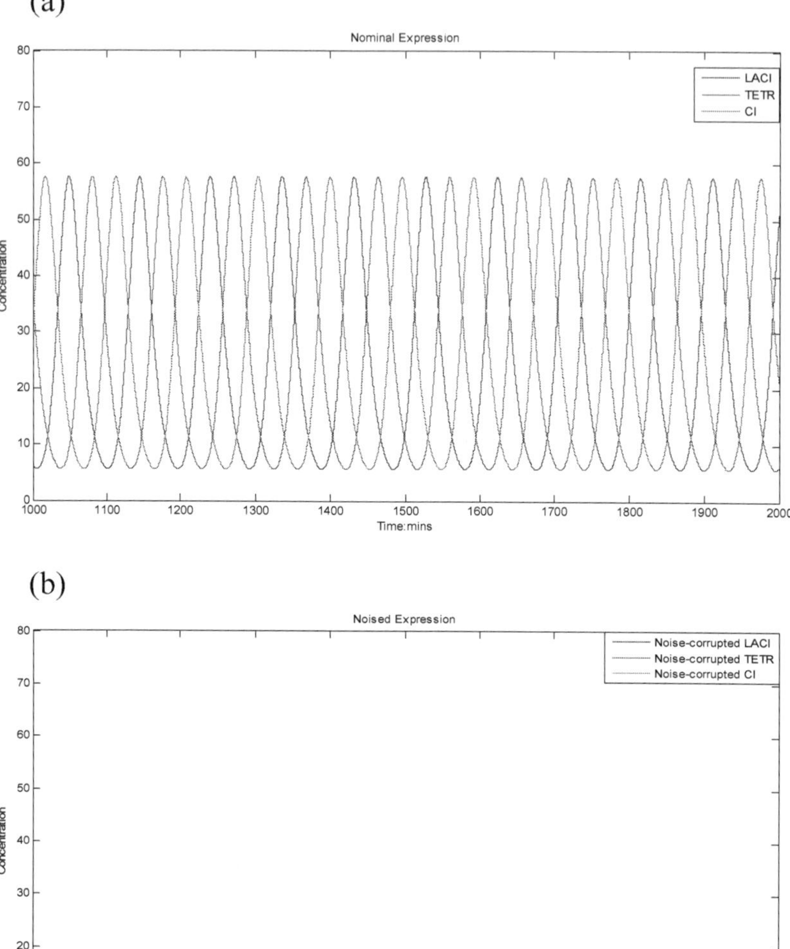

(b)

Figure 4.4. Time-responses of protein concentrations. (a) The nominal repressilator time-response with $\alpha_i = 0.5$, $\gamma_{pi} = 0.069$ for $i = 1,2,3$ by the minute in (Elowitz and Leibler 2000). (b) The repressilator time-response under intrinsic parameter fluctuations and extrinsic disturbances on the host cell. These two time-responses show that the repressilator in (Elowitz and Leibler 2000) suffers substantially from the effects of intrinsic parameter fluctuations and environmental noises on the host cell. Clearly, the corrupted repressilator does not have enough robustness to tolerate parameter fluctuations and extrinsic noises and loses its characteristics of oscillation.

Color image of this figure appears in the color plate section at the end of the book.

Suppose the synthetic gene network is affected by the four random intrinsic parameter fluctuations from n_1 to n_4, and environmental disturbances $v_1(t)$, $v_2(t)$ and $v_3(t)$ as follows

$$\frac{dm_1}{dt} = -\left(0.3465 + 0.15n_1\right)m_1 + \frac{\left(\alpha_1 + 0.75n_2\right)}{1+\left(p_3/40\right)^2} + \alpha_{01} + 0.1v_1$$

$$\frac{dp_1}{dt} = \left(0.167 + 0.1n_3\right)m_1 - \left(\gamma_{p_1} + 0.5n_4\right)p_1$$

$$\frac{dm_2}{dt} = -\left(0.3465 + 0.12n_1\right)m_2 + \frac{\left(\alpha_2 + 0.375n_2\right)}{1+\left(p_1/40\right)^2} + \alpha_{02}$$

$$\frac{dp_2}{dt} = \left(0.167 + 0.2n_3\right)m_2 - \left(\gamma_{p_2} + 0.1n_4\right)p_2 + v_2$$

$$\frac{dm_3}{dt} = -\left(0.3465 + 0.1n_1\right)m_3 + \frac{\left(\alpha_3 + 0.5n_2\right)}{1+\left(p_2/40\right)^2} + \alpha_{03}$$

$$\frac{dp_3}{dt} = \left(0.167 + 0.15n_3\right)m_3 - \left(\gamma_{p_3} + 0.1n_4\right)p_3 + v_3$$

$$(4.13)$$

where $\alpha_{0i} = 0.001\alpha_i$ for $i = 1, 2, 3$; and the random fluctuation sources n_1, n_2, n_3, and n_4 are independent standard white noises with unit variance from transcriptional noise, translational noise, and molecular diffusion noise, etc. The standard deviations of stochastic parametric fluctuations in the host cell are assumed as

$$\begin{bmatrix} \delta\gamma_{m1} & \delta\gamma_{m2} & \delta\gamma_{m3} \\ \delta\alpha_1 & \delta\alpha_2 & \delta\alpha_3 \\ \delta\beta_1 & \delta\beta_2 & \delta\beta_3 \\ \delta\gamma_{p_1} & \delta\gamma_{p_2} & \delta\gamma_{p_3} \end{bmatrix} = \begin{bmatrix} 0.15 & 0.12 & 0.1 \\ 0.75 & 0.375 & 0.5 \\ 0.1 & 0.2 & 0.15 \\ 0.5 & 0.1 & 0.1 \end{bmatrix}$$

and the external noises v_k for $k = 1, 2, 3$ are assumed to be uniformly distributed white noises on the concentrations of corresponding genes or proteins to show the ubiquitous intercellular and environmental disturbances.

Under the intrinsic fluctuations and extrinsic noises, the nominal and the noise-corrupted protein time-responses of a synthetic gene network with $\alpha_i = 0.5$ and $\gamma_{pi} = 0.069$ in (Elowitz and Leibler 2000) are shown in Figure 4.4a and Figure 4.4b, respectively. With intrinsic parameter perturbations and extrinsic disturbances, the parameter fluctuations of the repressilator can perturb the system away from the limit cycle region in (Elowitz and Leibler 2000), i.e., the bifurcation in parameters can perturb the synthetic genetic system away from limit cycle region and converge to some stable region with steady state values shown in Figure 4.4b. Obviously, the oscillatory character of synthetic repressilator in (Elowitz and Leibler 2000) has been violated by intrinsic parameter fluctuations and extrinsic noises. Therefore, the repressilator in (Elowitz and Leibler 2000) is not a robust oscillator. In this situation, a robust synthetic genetic oscillator design is necessary to guarantee the synthetic gene oscillator to function properly under intrinsic parameter fluctuations and extrinsic disturbances.

In order to solve the robust optimal tracking design problem of synthetic gene oscillator via the proposed GA-based design algorithm, we set the GA operators as follows (Chen and Chen 2010): first, we use the roulette wheel selection to increase the selecting efficiency of the populations which have higher fitness score; second, the crossover rate is 0.8; third, the chromosome mutates uniformly with the mutation rate 0.05; and fourth, we choose $[F_b \ F_w \ J_b \ J_w] = [10 \ 0.01 \ 5.5869 \ 2.6779]$. In the binary coding process, we set the bit length as $[B_{\alpha_1} \ B_{\alpha_2} \ B_{\alpha_3} \ B_{\gamma_{p1}} \ B_{\gamma_{p2}} \ B_{\gamma_{p3}}] = [11 \ 11 \ 11 \ 10 \ 10 \ 10]$, i.e., the corresponding resolutions of R_{α_i} and $R_{\gamma_{pi}}$ are specified as $[0.005 \ 0.005 \ 0.005 \ 0.004 \ 0.004 \ 0.004]$ among the feasible parameter ranges given by $\alpha_i \in [0 \ 10]$ and $\gamma_{pi} \in [0 \ 5]$ for $i = 1, 2, 3$ (in (Elowitz and Leibler 2000), $\alpha_i = 0.5$ and $\gamma_{pi} = 0.069$ for all i). Via the help of Genetic Algorithm Toolbox in Matlab, from these feasible parameter ranges, we could solve the optimal tracking problem of robust synthetic oscillator as $[\alpha_1^* \ \alpha_2^* \ \alpha_3^* \ \gamma_{p_1}^* \ \gamma_{p_2}^* \ \gamma_{p_3}^*] = [0.515 \ 0.505 \ 0.505 \ 0.068 \ 0.068 \ 0.072]$, with fitness score 10. During the GA simulation, we record the course of evolutionary history in Figure 4.5. From the simulation result, we can see that the proposed GA-based design method could find the optimal oscillation tracking solution efficiently. Because of the random process of mutation and crossover, the average fitness score jumps up and down. However, the best value is improved by employing the elite strategy with the best two populations in the evolution process. This can save the optimal solution until the crossover and mutation processes of the next generation. In Figure 4.5, we show both the evolutions of the best cost value and the best fitness score. During the evolutionary process, because the repressilator system is a dynamic synthetic gene network, i.e., the parameter fluctuations and extrinsic disturbances vary stochastically in each generation, the fitness score of the best population in the new generation would be slightly different from the fitness score of the same population in the old generation. This is why the fitness score in Figure 4.5 is unsteady.

Based on the design parameters via the proposed GA-based design method, the time responses of robust synthetic gene oscillator under intrinsic parameter fluctuations and extrinsic noises are shown in Figure 4.6. Through the robust GA-based design method, we can obtain desired oscillatory

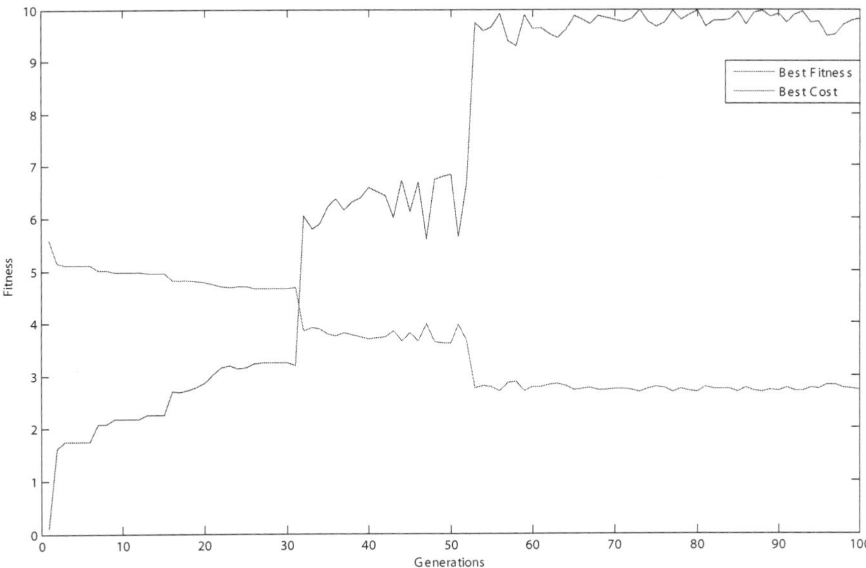

Figure 4.5. Convergence of fitness value. The best fitness score and best cost value evolve during the generations. The vibrations of the best cost value and the best fitness score come from the stochastic intrinsic parameter fluctuations and extrinsic noises, which fluctuate in each generation and directly affect the reliability of the synthetic gene network.

Color image of this figure appears in the color plate section at the end of the book.

behavior in this repressilator system. In Figure 4.6b, the repressilator system with GA optimal solutions shows the robust desired characteristics of oscillation under intrinsic parameter fluctuations and extrinsic disturbances. Although there are still some discrepancies between the desired oscillation signals and the protein concentrations of repressilator, mainly due to the intrinsic parameter fluctuations and extrinsic disturbances, these results are much better than the synthetic design in (Elowitz and Leibler 2000) as shown in Figure 4.4b. Clearly, the proposed robust synthetic gene oscillator design method has potential for practical applications in future.

In this *in silico* robust repressilator design example, the robust design scheme could be realized with the specified robust mRNA transcription rates, α_i^*, and protein decay rates, γ_{pi}^* for $i = 1,2,3$, in the corresponding feasible parameter design ranges to satisfy the prescribed oscillatory characteristics of the synthetic gene oscillator. Several biotechnology methods have been proposed by adjusting the combinatorial polymerase binding boxes and integrating different ligations to generate a diverse promoter library and a diverse protein decay rate (Basu et al. 2005, Hammer et al. 2006, Cox et al. 2007).Thus, we could synthesize the genetic repressilator with fine-tuned parameters, α_i and γ_{pi} for $i = 1,2,3$ in (4.10) to confirm our design scheme.

(a)

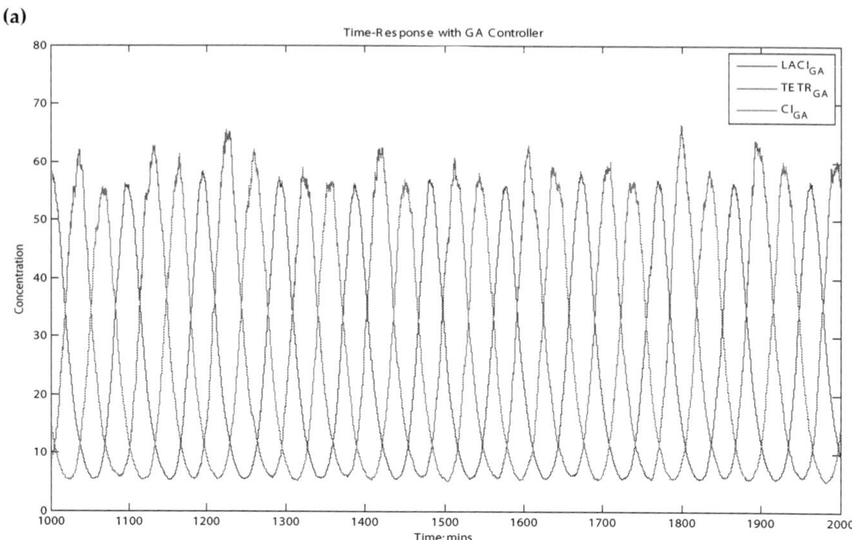

(b)

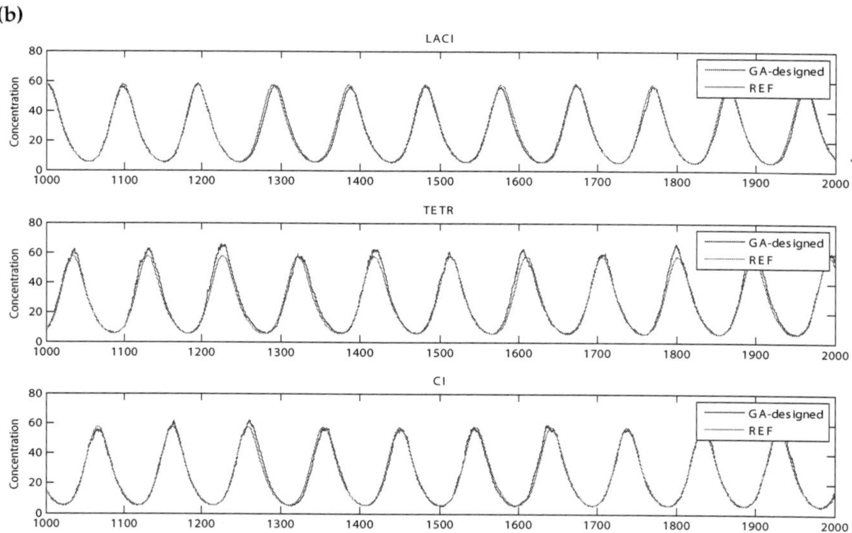

Figure 4.6. Time-response of the synthetic gene oscillator via the proposed GA-based design method solution. (a) Time-responses of these three proteins. (b) Time-response tracking of each protein and its reference. Under the parameter fluctuations and environmental noises, the designed repressilator can maintain its characteristics of oscillation and function properly. There are still some discrepancies between the desired reference signals and the protein concentrations of the repressilator, which are mainly due to parameter perturbations and environmental noises.

Color image of this figure appears in the color plate section at the end of the book.

Although this method is direct, it is an inefficient method. To remedy this, synthetic biologists can increase efficiency of synthetic gene network design through registries of biological parts and standard data sheets of the BioBrick assembly standard, which are developed and concerned with proper packing and characterizing of 'modular' biological activities such that these biological parts or devices with some prescribed characters may be efficiently assembled into gene networks in the future (Canton et al. 2008).

Comprehensive data sheets are used for quantitative descriptions of devices in many standardized engineering disciplines. A synthetic gene network designer can quickly and easily select some desired devices from biological device data sheets to meets their design requirements of a system (Canton et al. 2008). Thus, via the help of engineering theory and experience, a conceived system could be constructed by a set of devices with standard characteristics, which are typically reported on data sheets and are common across a wide range of devices type, such as sensors, logic elements and actuators. Biological data sheets have been set as standards for the characterization, manufacture and sharing of information on modular biological devices to promote a more efficient, predictable and design-driven genetic engineering science (Arkin 2008, Canton et al. 2008). Because data sheets of biological parts or devices embody engineering standards for synthetic biology, a good device standard should show sufficient information about biological parts or devices to allow the design of synthetic gene networks with optimal parameters. Data sheets contain a formal set of context-dependent, input-output behaviors, tolerances, requirements and other details about a particular biological part or device (Arkin 2008, Canton et al. 2008). Since parameters α_i represent the optimal transcriptional rates, these parameters could be measured from input-output behaviors of biological parts or devices. Moreover, through the combinations of one or more devices, a biological designer could assemble another well-defined physical device, such as a well-defined γ_{pi} to achieve a desired oscillatory characteristic. From properly characterized input-output behaviors, the biological designer can estimate the corresponding parameters of biological parts or devices. Then, adequate devices can be rapidly selected from an extensive list of the parts to meet the design parameters. Adherence to the set of standard devices in data sheets ensures that each device and systems synthesized from them can satisfy the requirements of design specifications for a robust synthetic gene oscillator.

But there are many uncertainties about the behavior of synthetic oscillators. For example, the cellular functions from devices will fluctuate and there are also many unpredictable uncertainties among components as well as on the host cell. Since the transcriptional rate, α_i, has a positive correlation with the promoter activity, it can be seen as a combinative presentation of promoter strength and ribosome binding site of the transcription. However, there are still some variations or uncertainties on the parameter value α_i. In terms of a mathematic model, these variations or uncertainties of α_i can be transformed to an equivalent uncertain disturbance $\delta\alpha_i$ in (4.10), as can the other parameters $\delta\gamma_{mi}$, $\delta\beta_i$, and $\delta\gamma_{pi}$. The robust synthetic oscillator design can predict the most robust values of α_i and γ_{pi} by the proposed GA-based design method under intrinsic fluctuations and extrinsic noise. In the design example, the proposed synthetic gene oscillator not only can achieve the desired oscillation tracking design but also can tolerate the worst-case effect due to the seuncertain parameter fluctuations and external noises on the host cell.

By using the GA-based design method along with Matlab, we can easily solve the design parameters for this optimal reference tracking problem of a robust synthetic gene oscillator under intrinsic and extrinsic noises. However, there are still some disadvantages in the GA method. First, this method requires a great deal of time for the coding and decoding process in the natural selection if the number of design parameters increases. Fortunately, there are many advanced GA methods, like Hybrid Genetic Algorithm (HGA) (Renders and Flasse 1996, Katayama et al. 2000, Katayama and Narihisa 2001) or the combination with Simulated Annealing Algorithm (SA) (Kirkpatrick 1984, Khan et al. 1997), that can save time and increase the probability of finding the global optimum solution. Second, the solution may be only a near-optimum due to limitations of GA method, for example, the limitation of finite bit length B_i of chromosome coding and the finite generations in GA searching process. Therefore it is perhaps not surprising that the GA method may not converge to the truly global optimal tracking solution.

Despite these disadvantages of GA methods, their primary advantage is that the highly nonlinear constrained minimization problem can be solved for a robust synthetic gene oscillator, which does not have a closed-form solution. To avoid finding a local optimal solution, the proposed GA-based design method can help approach the global optimal solution by the 'mutation' and 'crossover' processes. Even though the GA method does not always find the global optimal solution, its solution is often close to the optimum, whereas other conventional searching algorithm can only obtain a local optimal solution. By the property of mimicking natural selection in the GA method, most optimal solutions are not reproducible in the repeated biological simulations. This is not surprising because the GA searching process contains not only the different initial conditions but also the different random mutations and crossovers, as in real world evolutionary processes. For example, in some *in vivo* experimental studies, *E. coli* changes its genotype to increase the survival opportunity when suffering intrinsic fluctuations and environmental noises like glucose fluctuation, glucose limitation, molecular thermal fluctuation or other environmental stresses (Papadopoulos et al. 1999, Cooper et al. 2003, Pelosi et al. 2006). These random mutations and crossovers may be different in the GA searching process, but the evolutionary results from the GA method would be very similar due to mimicking the natural selection to maximize the fitness score. This phenomenon is a form of convergent evolution and, clearly, the results by the GA-based design method *in silico* mirror what happens in these *in vivo* studies in parallel *E. coli* cultures.

From an engineering point of view, when we synthesize a prescribed biological oscillator as the repressilator, its function could suffer interference from the intrinsic fluctuations and environmental noises that affect the host cell. These fluctuations and noises will corrupt the synthetic gene oscillator so that it cannot achieve the desired behavior. In this chapter, we proposed a design procedure using the GA method which mimics the natural selection in the evolutionary process of the real world to optimize the desired reference tracking of synthetic gene oscillator and to tolerate parameter fluctuations and external disturbances on the host cell. In this respect, our design is a rapid selection scheme. This can save the evolution time for optimal selection in the revolutionary process for increasing the robust oscillation characteristics and for improving the reliability of a synthetic gene network. Thus, the time responses in Figure 4.6b compared with the time response in Figure 4.4b show that the robustly designed repressilator can efficiently eliminate the effect of uncertainties due to effects of intrinsic parameter fluctuation and the extrinsic noise on the oscillation.

Clearly, the proposed GA-based design method provides a systematic design method for a robust synthetic gene oscillator with desired amplitude, frequency and phase in a host cell with intrinsic parameter fluctuations and external disturbances. Therefore, combined with the recently advanced synthetic techniques such as promoter library, ssrA-tagged protein or the BioBrick assembly standard devices in biological device data sheets, the proposed design method has good potential for practical applications of robust synthetic genetic oscillators in future.

Recently, the synchronization problems of coupled biochemical oscillations have been widely studied (Wang and Chen 2005, Zhou et al. 2005, Wang et al. 2006, Zhou et al. 2008). This is an important topic of synthetic gene oscillators for practical applications. Therefore, the robust synchronization design problem of a large number of coupling synthetic gene oscillators under intrinsic fluctuations and external disturbances will be discussed in Chapter 8.

4.3 An *in silico* Design Example: EA Approach

Consider the biological AND gate as shown in Figure 2.4 and the dynamic differential equations of biological AND gate in (2.5) (Terzer et al. 2007). From the nonlinear differential equations as mentioned above, it can be seen that the dynamics of the biological AND gate depends on some biochemical factors, such as the kinetic constant, degradation constant and basal level. However, these factors or parameters are uncertain inherently and the biological circuit also suffers from environmental noises. In this situation, the dynamic model of synthetic biological circuit *in vivo* should be modified as follows (Chen et al. 2011).

$$\frac{dx_{mT7Pol}}{dt} = (k_1 + \Delta k_1 n_1)u_1 - (\lambda_1 + \Delta \lambda_1 n_4)x_{mT7Pol} + v_1$$

$$\frac{dx_{mtRNA}}{dt} = (k_2 + \Delta k_2 n_1)u_2 - (\lambda_2 + \Delta \lambda_2 n_4)x_{mtRNA} + v_2$$

$$\frac{dx_{T7Pol^*}}{dt} = (k_3 + \Delta k_3 n_2)x_{mT7Pol} - (\lambda_3 + \Delta \lambda_3 n_4)x_{T7Pol^*} + v_3$$

$$\frac{dx_{tRNA}}{dt} = (k_4 + \Delta k_4 n_2)x_{mtRNA} - (\lambda_4 + \Delta \lambda_4 n_4)x_{tRNA} + v_4$$

$$\frac{dx_{T7Pol}}{dt} = (k_5 + \Delta k_5 n_3)x_{mT7Pol}x_{tRNA} - (\lambda_5 + \Delta \lambda_5 n_4)x_{T7Pol} + v_5$$

$$\frac{dx_{mGFP}}{dt} = P_{mGFP} + \frac{(k_6 + \Delta k_6 n_1)\left(\frac{x_{T7Pol}}{K}\right)^n}{1 + \left(\frac{x_{T7Pol}}{K}\right)^n} - (\lambda_6 + \Delta \lambda_6 n_4)x_{mGFP} + v_6$$

$$\frac{dx_{GFP}}{dt} = (k_7 + \Delta k_7 n_2)x_{mGFP} - (\lambda_7 + \Delta \lambda_7 n_4)x_{GFP} + v_7$$

$$(4.14)$$

where Δk_i, $\Delta \lambda_i$ denote the amplitudes of parameter fluctuations for kinetic constants and degradation constant, respectively. n_j are random white noises with zero mean and unit variance, which denote the independent random fluctuation sources in transcription, translation, reaction and degradation process, respectively. The variances of parameter perturbations are given as $\mathrm{var}(\Delta k_i n_j) = (\Delta k_i)^2$, $\mathrm{var}(\Delta \lambda_i n_j) = (\Delta \lambda_i)^2$, i.e., Δk_i and $\Delta \lambda_i$ represent the standard deviations of the corresponding stochastic parameter fluctuations to be tolerated by the synthetic gene circuit *in vivo*. v_i denotes the corresponding external noise with variance σ_i^2. Note that the perturbation of basal level P_{mGFP} is merged into v_6.

The design specifications and parameters are given as follows. The desired transient behaviors of biological AND gate are as shown in Figure 4.7. The standard deviations of uncertain fluctuations of kinetic parameters and decay rates to be robustly tolerated are specified as $\Delta k_1 = 0.2$, $\Delta k_2 = 0.2$, $\Delta k_3 = 0.02$, $\Delta k_4 = 0.02$, $\Delta k_5 = 0.000005$, $\Delta k_6 = 0.2$, $\Delta k_7 = 0.02$, $\Delta \lambda_1 = 0.04$, $\Delta \lambda_2 = 0.04$, $\Delta \lambda_3 = 0.05$, $\Delta \lambda_4 = 0.05$, $\Delta \lambda_5 = 0.05$, $\Delta \lambda_6 = 0.04$ and $\Delta \lambda_7 = 0.05$. The variance of external disturbance v_i is $\sigma_i^2 = (0.1)^2$. The feasible ranges of kinetic parameters to be designed are specified in the range from 0 to 1. The other parameters are set as $\lambda_1 = \lambda_2 = \lambda_6 = 0.0231$, $\lambda_3 = \lambda_5 = \lambda_7 = 0.0023$, $\lambda_4 = 0.0002$, $K = 400$ and $n = 1$ (Terzer et al. 2007).

Given the above design specifications, our design objective is to adapt the design parameters by the proposed network evolutionary method to achieve the optimal tracking under intrinsic parametric fluctuations and

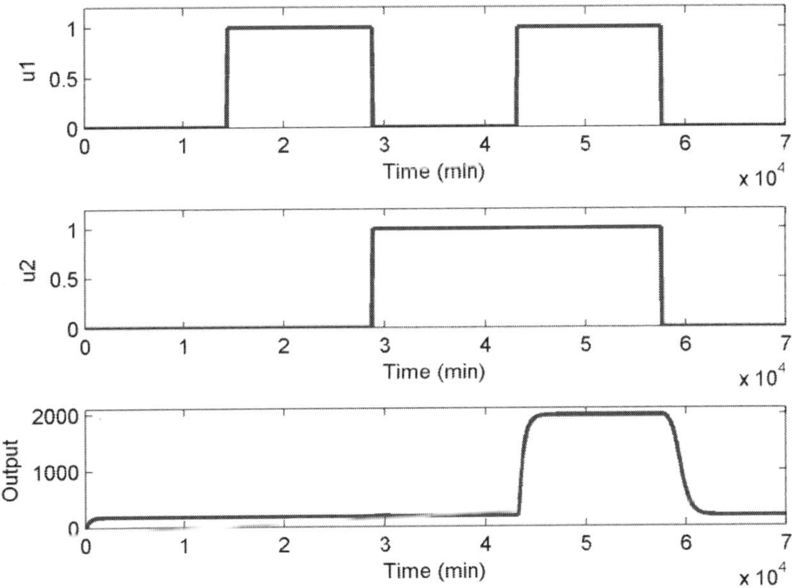

Figure 4.7. Desired transient behaviors of biological AND gate. The biological AND gate generates an output signal only when it gets biochemical signals from both of its inputs.

extrinsic noises. In this example, the parameters of evolutionary algorithm are chosen as $M=100$, $p_c=0.9$ and $p_m=0.2$. The software MATLAB is used to perform the simulation. After 100 generations, the best fit design parameters $k_1^* = 0.6042$, $k_2^* = 0.8410$, $k_3^* = 0.9272$, $k_4^* = 0.2063$, $k_5^* = 0.00001$, $k_6^* = 0.8640$ and $k_7^* = 0.1235$ are obtained by the proposed network evolutionary method with the best fitness $F(k^*) = 1.09 \times 10^{-4}$. By Monte Carlo simulation with 50 rounds, the output of the synthetic gene network with the design parameters under intrinsic parametric fluctuations and extrinsic noises is shown in Figure 4.8a. It can be seen that the synthetic gene network has robust regulation ability to achieve the desired transient behaviors in spite of uncertain initial state, kinetic parameter fluctuations and disturbances on the host cell. On the contrary, as shown in Figure 4.8b, if the design parameters are selected aside the optimal design parameter k^*, for example, with $k_1 = 0.9$, $k_2 = 0.6$, $k_3 = 0.95$, $k_4 = 0.25$, $k_5 = 0.00002$, $k_6 = 0.6$ and $k_7 = 0.15$ and the fitness $F(k) = 3.54 \times 10^{-5}$, the expression of the synthetic gene network is with more fluctuation and cannot achieve the desired transient behaviors. Obviously, the robust synthetic gene network by the proposed network evolutionary method has a good robust stability to overcome the uncertain initial conditions and an enough filtering ability to attenuate the disturbances on the host cell and eventually approach the desired transient behaviors.

4.4 Summary

In the light of natural selection on traits best suited for environmental change being an important mechanism in evolution, the similar genetic or evolutionary strategy seems more suitable to gene circuit design if we can speed up the evolutionary process through fast parallel evolutionary computations. Inspired by biological evolution such as reproduction, mutation, recombination, and selection, the genetic and evolutionary algorithms are efficient methods to solve optimization problems for systematic circuit design in synthetic biology. A network evolutionary algorithm is employed here to solve the best fitness function to gradually improve the desired behavior tracking ability of a synthetic gene circuit design one generation by one generation to mimic the natural selection in the evolutionary process. Unlike the necessity of some complicated computations in the conventional design strategies, only some simple operators (e.g., selection, crossover, and mutation) and some simple calculations are required for selecting optimal design parameters iteratively by the proposed network evolutionary design method, but with synthetic gene circuit robust enough against intrinsic parameter fluctuation and external disturbance. This attractive property makes the proposed network

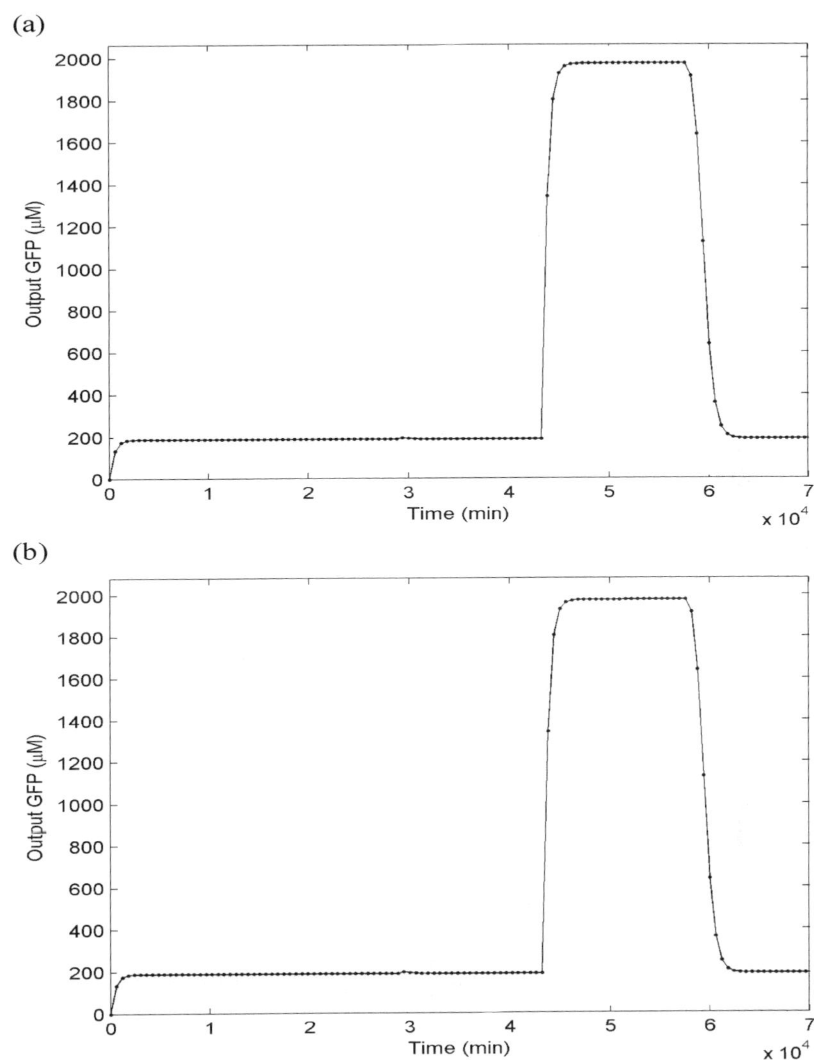

Figure 4.8. Simulation results of biological AND gate. (a) The simulation result with optimal design parameters k_1^*=0.6042, k_2^*=0.8410, k_3^*=0.9272, k_4^*=0.2063, k_5^*=0.00001, k_6^*=0.8640 and k_7^*=0.1235. The Monte Carlo simulations are performed by 50 rounds. The mean error is

$$\bar{e} = \frac{1}{50}\sum_{i=1}^{50} e_i = 121.95$$ with standard deviation of 9.93, where e_i is the root mean squared error of the ith Monte Carlo simulation. (b) In contrast to the above design case, the design parameters are specified aside the optimal design parameter k^*, with k_1=0.9, k_2=0.6, k_3=0.95, k_4=0.25, k_5=0.00002, k_6=0.6 and k_7=0.15. In this design case, the mean error is $\bar{e} = 165.90$ with standard deviation of 18.67.

Color image of this figure appears in the color plate section at the end of the book.

evolutionary method being easily implemented in the practical applications. Since the proposed network evolutionary design method has included design specifications such as the tolerable variances of intrinsic stochastic parameter variations and external disturbances, the feasible ranges of design kinetic parameters and decay rates, and the desired steady state or transient behaviors of output $y(t)$, the synthetic gene circuit can be guaranteed to achieve all possible design purposes by only solving a constrained optimization problem in (4.2) by maximizing a corresponding fitness function in (4.4) through EA or GA algorithm. Though the proposed evolutionary gene circuit designs may not produce the best tracking performance within the finite generations of EA or GA algorithm, they are near optimal gene circuit designs to achieve the desired behavior. Therefore, the proposed network evolutionary method has potential applications to the synthetic gene circuit design for biotechnological purposes in the near future.

References

Arkin, A. 2008. Setting the standard in synthetic biology. Nat Biotechnol 26: 771–774.

Basu, S., Gerchman, Y., Collins, C.H., Arnold, F.H. and Weiss, R. 2005. A synthetic multicellular system for programmed pattern formation. Nature 434: 1130–1134.

Canton, B., Labno, A. and Endy, D. 2008. Refinement and standardization of synthetic biological parts and devices. Nature biotechnology 26: 787–794.

Chen, B.S. and Chen, P.W. 2010. GA-based Design Algorithms for the Robust Synthetic Genetic Oscillators with Prescribed Amplitude, Period and Phase. Gene Regul Syst Bio 4: 35–52.

Chen, B.S., Hsu, C.Y. and Liou, J.J. 2011. Robust design of biological circuits: evolutionary systems biology approach. J Biomed Biotechnol 2011: 304236.

Cooper, T.F., Rozen, D.E. and Lenski, R.E. 2003. Parallel changes in gene expression after 20,000 generations of evolution in *Escherichia coli*. Proceedings of the National Academy of Sciences 100: 1072–1077.

Cox, R.S., 3rd, Surette, M.G. and Elowitz, M.B. 2007. Programming gene expression with combinatorial promoters. Mol Syst Biol 3: 145.

Elowitz, M.B. and Leibler, S. 2000. A synthetic oscillatory network of transcriptional regulators. Nature 403: 335–338.

Hammer, K., Mijakovic, I. and Jensen, P.R. 2006. Synthetic promoter libraries—tuning of gene expression. Trends Biotechnol 24: 53–55.

Katayama, K., Sakamoto, H. and Narihisa, H. 2000. The efficiency of hybrid mutation genetic algorithm for the travelling salesman problem. Mathematical and computer modelling 31: 197–203.

Katayama, K. and Narihisa, H. 2001. An Efficient Hybrid Genetic Algorithm for the Traveling Salesman Problem. Electronics & Communications in Japan, Part III: Fundamental Electronic Science (English translation of Denshi Tsushin Gakkai Ronbunshi) 84: 76–83.

Khan, Z., Prasad, B. and Singh, T. 1997. Machining condition optimization by genetic algorithms and simulated annealing. Computers and Operations Research 24: 647–657.

Kirkpatrick, S. 1984. Optimization by simulated annealing: Quantitative studies. Journal of Statistical Physics 34: 975–986.

Papadopoulos, D., Schneider, D., Meier-Eiss, J., Arber, W., Lenski, R.E. and Blot, M. 1999. Genomic evolution during a 10,000-generation experiment with bacteria. National Acad Sciences, 3807–3812.

Pelosi, L., Kuhn, L., Guetta, D., Garin, J., Geiselmann, J., Lenski, R.E. and Schneider, D. 2006. Parallel Changes in Global Protein Profiles During Long-Term Experimental Evolution in Escherichia coli. Genetics 173: 1851.

Renders, J.M. and Flasse, S.P. 1996. Hybrid methods using genetic algorithms for global optimization. Systems, Man and Cybernetics, Part B, IEEE Transactions on 26: 243–258.

Terzer, M., Jovanovic, M., Choutko, A., Nikolayeva, O., Korn, A., Brockhoff, D., Zurcher, F., Friedmann, M., Schutz, R., Zitzler, E., Stelling, J. and Panke, S. 2007. Design of a biological half adder. Synthetic Biology, IET 1: 53–58.

Wang, R. and Chen, L. 2005. Synchronizing genetic oscillators by signaling molecules. Journal of biological Rhythms 20: 257.

Wang, R., Chen, L. and Aihara, K. 2006. Synchronizing a multicellular system by external input: an artificial control strategy. Bioinformatics 22: 1775–1781.

Zhou, T., Chen, L. and Aihara, K. 2005. Molecular communication through stochastic synchronization induced by extracellular fluctuations. Physical review letters 95: 178103.

Zhou, T., Zhang, J., Yuan, Z. and Chen, L. 2008. Synchronization of genetic oscillators. Chaos: An Interdisciplinary Journal of Nonlinear Science 18: 037126.

5

Construction of Promoter and Promoter-RBS Libraries for Synthesis of Gene Networks

In previous chapter, some design methods have been developed to select some well-matched biological parts to combine and produce the desired behavior reliably. However, the above design methods need to tune some kinetic parameters to achieve some desired steady or oscillatory state. In fact, tuning intrinsic parameters of biological parts to fit the designed parameter is currently quite a difficult or even unfeasible task for biotechnology. Further, it is still hard to select adequate biological parts to implement a desired cellular function with quantitative value. With this concept, we think that selecting adequate components from the existing component libraries for synthetic gene networks is more convenient than tuning parameters of some components to achieve the designed values of synthetic gene networks. To overcome this problem, synthetic biologists usually create many versions of synthetic parts with diverse characteristics by direct evolution, point mutation or random combination of DNA components, and the functions of these versions are investigated to engineer the gene network with desired behavior. But when the design of gene network is complex, the numbers of biological parts needed to be created and tested is dramatically increased. At present, some well-characterized promoter libraries for engineering gene networks are widely available. Thus a synthetic gene network can be constructed by selecting adequate promoters from promoter libraries to achieve the desired behaviors. However, the present promoter libraries only have information of the maximum and minimum gene expressions of biological parts and cannot be directly applied to engineer a synthetic gene network. In order to efficiently select adequate promoters from promoter libraries for synthetic gene network, promoter libraries are needed to be redefined based on the dynamic gene regulation. In addition, promoters and ribosome binding sites (RBSs) are regarded as a lumped component for synthetic gene network design. How to construct the promoter-RBS libraries are also discussed in this chapter.

5.1 Construction of Promoter Library

Quantitative fluorescence measurements of promoters have been used to build promoter libraries over the past decade (Alper et al. 2005, Hammer et al. 2006, Cox et al. 2007, Murphy et al. 2007). The prospect of engineering a synthetic gene network with desired behaviors by selecting adequate promoters from these existing promoter libraries is appealing for synthetic biologists. However, the current promoter libraries provide only the experimental output of fluorescence as an indirect measure of promoters and still lack of information on the promoter activities, which are useful for engineering synthetic gene networks. At present, there is still not an efficient method to quickly select adequate promoters from the current promoter libraries for synthetic gene networks.

In general, a synthetic gene network always consists of a set of promoter-regulation circuits as shown in Figure 5.1. Therefore, the dynamic model of promoter-regulation gene circuit in Figure 5.1 should be studied based on promoter activities in promoter libraries before the design of synthetic gene network. Since directly measuring promoter activities is still challenging, measuring fluorescence as an indirect measure of promoter activities is widely used (Murphy et al. 2007, Rosenfeld et al. 2007, Canton et al. 2008, Ellis et al. 2009, Kelly et al. 2009). Nowadays the current promoter libraries only contain the fluorescence of promoter. However, the promoter activities rather than fluorescence are directly employed as design parameters to

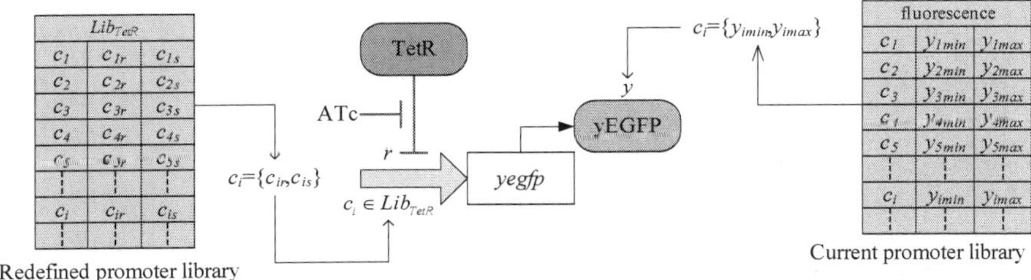

Figure 5.1. Single schematic diagram of the synthetic promoter-regulation gene circuit.

select adequate promoters in engineering a synthetic gene network. For the convenience of selection, we redefine previous promoter libraries in the following.

A dynamic model can be used to indirectly evaluate the promoter activities (Leveau and Lindow 2001, Ellis et al. 2009, Kelly et al. 2009) to help redefine existing promoter libraries shown in Figure 5.1. Suppose the promoter c is selected from the TetR-regulated promoter library. In general, the promoter c always has the minimum expression with the repressor TetR binding and the maximum expression without it. Furthermore, the repressor TetR is also regulated by the inducer adhydrotetracycline (ATc). Hence we denote a repressor activity r, which is regulated by the inducer ATc, of the repressor TetR with the following form (Alon 2007):

$$r = \frac{TetR}{1 + (ATc / K_{ATc})^{n_{ATc}}} \tag{5.1}$$

where K_{ATc} denotes the ATc-TetR dissociation rate and n_{ATc} denotes the binding cooperativity between ATc and TetR. Then the promoter regulation function $p_{TetR}(c,r)$, which is dependent on the repressor activity r and the choice of the promoter c, has the following form (Wu et al. 2011a, Wu et al. 2011b):

$$p_{TetR}(c,r) = c_r + \frac{c_s - c_r}{1 + (r / K_{TetR})^{n_{TetR}}} = c_r + (c_s - c_r)H_{TetR}(r)$$

$$c = \{c_r, c_s\} \in Lib_{TetR} \tag{5.2}$$

where the promoter c has two promoter activities c_r and c_s for the TetR-regulated promoter library Lib_{TetR}, i.e., $c = \{c_r, c_s\} \in Lib_{TetR}$; and K_{TetR} and n_{TetR} denote the TetR-DNA binding affinity and binding cooperativity between regulatory protein TetR and DNA, respectively; $H_{TetR}(r) = 1/(1 + (r / K_{TetR})^{n_{TetR}})$ is a Hill function which captures the regulatory effect of a regulatory protein (Alon 2007). According to the above definition, the promoter regulation function $p_{TetR}(c,r)$ in (5.2) has the minimum value c_r and maximum value c_s with and without the saturating concentration of the repressor activity in (5.1). These are related to the minimum fluorescence y_{min} and maximum fluorescence y_{max} of fluorescent protein. By the simple estimation from fluorescence of fluorescent protein, c_r and c_s can simply transformed through y_{min} and y_{max}.

Here, we provide a method to redefine promoter libraries based on the estimated promoter activities from experimental data on the current promoter libraries in Ellis (Ellis et al. 2009). Although the experimental data of TetR- and LacI-regulated promoter libraries have already been reported (Ellis et al. 2009), it is necessary to have an efficient method of selecting promoters from promoter libraries to engineer a dynamic synthetic gene network in (5.2). Hence the promoter libraries need to be reconstructed by the dynamic model of promoter-regulation gene circuit in (5.2). Since TetR- and LacI-regulated promoter libraries are characterized similarly, only the redefinition of TetR-regulated promoter library needs to be described.

A simple scheme of Ellis' experimental model is shown in Figure 5.1. The constitutive expression of TetR represses *yegfp* by inhibiting the promoter *c* (TetR-regulated promoter) and then ensures low basal levels of yEGFP. On the other hand, TetR is repressed first by the saturating concentration of the inducer ATc and then by the promoter *c* activities so the gene *yegfp* transcripts and translates to ensure the high levels of yEGFP. Therefore, the maximum and minimum values of yEGFP are measured with and without the saturating concentrations of the inducer ATc, respectively. Based on these experimental data, we estimate the corresponding promoter activities and redefine these two promoter libraries by their promoter activities. The experimental model for estimating promoter activity in Figure 5.1 can be described as the following dynamic model (Leveau and Lindow 2001, Kelly et al. 2009, Wu et al. 2011a, Wu et al. 2011b)

$$\begin{cases} \dot{x}(c,t) = p_{TetR}(c,r) - \beta x(c,t) \\ \dot{X}(c,t) = \alpha x(c,t) - \gamma_{yEGFP} X(c,t) \end{cases} \tag{5.3}$$

$$y(c,t) = \eta X(c,t)$$

where β and γ_{yEGFP} denote the degradation rates of mRNA and protein yEGFP, respectively; α denotes the translation rate; $y(c,t)$ denotes the fluorescence of the protein yEGFP; η is the ratio of fluorescence to the concentration of yEGFP (Leveau and Lindow 2001). The promoter regulation function $p_{TetR}(c,r)$ in (5.2) and (5.3) has the maximum value c_s and minimum value c_r with and without saturating concentrations of the inducer ATc, respectively. This is because, with the saturating concentration of inducer ATc (i.e., the value of ATc is much larger than K_{ATc}), the repressor activity r in (5.1) becomes much smaller than K_{TetR} in (5.2) and the maximum of $p_{TetR}(c,r)$, $p_{TetR}^{max}(c,r) \approx c_r + c_s - c_r = c_s$ in (5.2). Furthermore, in the case without saturation of the inducer ATc, r in (5.1) becomes much larger than K_{TetR} in (5.2) so that the minimum of $p_{TetR}(c,r)$, $p_{TetR}^{min}(c,r) \approx c_r$ in (5.2). Therefore, in the case of saturation of inducer ATc, the steady state of $y(c,t)$ will have a maximum value, i.e.,

$$y_{max}(c,t) \approx \frac{\alpha}{\beta} \frac{\eta}{\gamma_{yEGFP}} p_{TetR}^{max}(c,r) = \frac{\alpha}{\beta} \frac{\eta}{\gamma_{yEGFP}} c_s \tag{5.4}$$

Similarly, in the case without the inducer ATc, the steady state of $y(c,t)$ will have a minimum value, i.e.,

$$y_{\min}(c,t) \approx \frac{\alpha}{\beta} \frac{\eta}{\gamma_{yEGFP}} p_{TetR}^{\min}(c,r) = \frac{\alpha}{\beta} \frac{\eta}{\gamma_{yEGFP}} c_r \tag{5.5}$$

Since the maximum and minimum values of $y_{\max}(c,t)$ and $y_{\min}(c,t)$ are both at steady states of experimental data in the TetR-regulated promoter library (Ellis et al. 2009) and the parameters α, β, η, γ_{yEGFP} can also be obtained from the experimental data of TetR-regulated promoter library (Ellis et al. 2009), the promoter activities c_r and c_s of every promoter in the promoter libraries can be estimated by these experimental data from (5.4) and (5.5), as listed in Table 5.1 (Wu et al. 2011a, Wu et al. 2011b). Other parameters to be needed for the above promoter activity estimation process are identified by least square parameter identification method in the following.

The repressor activity r in (5.1) becomes much larger or smaller than K_{TetR} with or without the inducer ATc and will lead to the promoter regulation functions $p_{TetR}^{\min}(c,r)$ and $p_{TetR}^{\max}(c,r)$, respectively. We need to identify K_{TetR} and K_{ATc} to determine the promoter activity of promoter regulation function $p_{TetR}(c,r)$. By employing Ellis' experimental data, in which the inducer ATc is the input and the normalized fluorescence is the output, concentrations of the inducer ATc in (5.1) and the promoter regulation function in (5.2) are used to identify two parameters K_{TetR} and K_{ATc} by the least square parameter identification method. The identified results are listed in Table 5.2 (Wu et al. 2011a, Wu et al. 2011b). By a similar process, the promoter activity $c = \{c_r, c_s\}$ of every promoter in the LacI-regulated promoter library can be obtained in Table 5.1 and two parameters K_{LacI} and K_{IPTG} can be identified by the least square parameter estimated method in Table 5.2. The other promoters from the corresponding promoter libraries could be similarly redefined based on the promoter activity. Since a complex synthetic gene network always consists of a set of promoter-regulation gene circuit in Figure 5.1, the design of the complex synthetic gene network becomes how to select a set of promoters from the corresponding promoter libraries with adequate promoter activities to achieve the design specifications. The detailed design procedure will be given in Chapter 6.

Table 5.1. Redefined constitutive, TetR- and LacI-regulated promoter libraries. The redefined constitutive, TetR- and LacI-regulated promoter libraries (i.e., Lib_{const}, Lib_{TetR} and Lib_{LacI}) comprise different promoters (i.e., J_k, $k=0, ..., 18$, T_k and L_k, $k=0, ..., 20$) with their corresponding promoter activities of c_s and c_r obtained from experimental data of previous promoter libraries (Ellis et al. 2009).

TetR-regulated promoter library (Lib_{TetR})			LacI-regulated promoter library (Lib_{LacI})			Constitutive promoter library (Lib_{const})		
Promoter	Promoter activity		Promoter	Promoter activity		Promoter	Promoter activity	
	c_s	c_r		c_s	c_r			c_s
T_0	2121	0.1724	L_0	1657.5	0.3018	J_0		5901.40
T_1	1604	0.7576	L_1	923.97	0.2567	J_1		5048.74
T_2	1376.6	0.1936	L_2	860.87	0.2244	J_2		4242.43
T_3	1169.8	0.4672	L_3	674.92	1.9189	J_3		4149.75
T_4	974.52	0.0753	L_4	651.58	1.1680	J_4		3445.38
T_5	942.77	0.2281	L_5	570.07	3.5062	J_5		3310.99
T_6	967.17	0.1493	L_6	527.83	0.5497	J_6		3019.05
T_7	738.57	0.0702	L_7	323.45	0.1248	J_7		2745.64
T_8	641.74	0.7135	L_8	327.77	0.1772	J_8		2103.84
T_9	564.24	0.2620	L_9	309.74	0.5439	J_9		1955.55
T_{10}	501.35	0.0756	L_{10}	298.35	0.1146	J_{10}		1443.49
T_{11}	469.35	0.0788	L_{11}	250.16	0.1326	J_{11}		917.53
T_{12}	466.16	0.1636	L_{12}	248.03	0.1171	J_{12}		896.68
T_{13}	356.88	0.0927	L_{13}	239.32	0.1010	J_{13}		593.15
T_{14}	348.95	0.1483	L_{14}	190.2	0.0959	J_{14}		375.35
T_{15}	274.79	0.1067	L_{15}	163.84	0.4813	J_{15}		245.60
T_{16}	250.04	0.0857	L_{16}	166.42	0.0989	J_{16}		48.66
T_{17}	188.77	0.1366	L_{17}	131.63	0.1190	J_{17}		39.39
T_{18}	119.57	0.0753	L_{18}	108.96	0.0903	J_{18}		2.32
T_{19}	111.57	0.1185	L_{19}	101.89	0.0982			
T_{20}	70.909	0.1606	L_{20}	85.673	0.2174			

Table 5.2. Mathematical characteristics of promoters and their regulation parameters. Parameter values are obtained from empirical studies in the literature or estimated via experimental data from Ellis et al. 2009. The parameters K_{TetR}, K_{LacI}, K_{ATc} and K_{IPTG} are identified from Ellis et al. 2009.

Parameter	Description	Value	Units	Ref.
K_{TetR}	TetR binding affinity	7.3093	M	+
K_{LacI}	LacI binding affinity	60.1405	M	+
K_{ATc}	ATc-TetR dissociation rate	26.3236	ng/ml	+
K_{IPTG}	IPTG-LacI dissociation rate	0.0598	mM	+
n_{TetR}	Binding cooperativity between TetR and DNA	2	--	(Braun et al. 2005)
n_{LacI}	Binding cooperativity between LacI and DNA	2	--	(Iadevaia and Mantzaris 2006)
n_{ATc}	Binding cooperativity between ATc and TetR	4	--	(Braun et al. 2005)
n_{IPTG}	Binding cooperativity between IPTG and LacI	1	--	(Iadevaia and Mantzaris 2006)
γ_{yEGFP}	yEGFP degradation rate	1.925×10^{-3}	min^{-1}	(Ellis et al. 2009)
γ_{TetR} and γ_{LacI}	TetR and LacI degradation rates	0.1386	min^{-1}	(Tuttle et al. 2005)
γ_{CI}	CI degradation rate	0.042	min^{-1}	(Arkin et al. 1998)
β	mRNA degradation rate	0.288	min^{-1}	(Canton et al. 2008, Kelly et al. 2009)
α	Translation rate	24	min^{-1}	(Canton et al. 2008, Kelly et al. 2009)

+: the parameters can be identified based on the experimental data.

5.2 Construction of Promoter-RBS Library

An exchangeable protocol of BioBrick-related data has been developed for a standardized, extensive, scalable and machine-processable interface (Canton et al. 2008). However, only few BioBrick parts are well-characterized. In fact, more well-characterized BioBrick parts are efficiently used in the synthetic gene circuit through systematic design. In order to facilitate the design of the synthetic gene circuit, the well-characterized BioBrick parts need to be constructed as BioBrick libraries. Since gene expression is both regulated by the promoter and RBS, we consider the promoter and RBS as a promoter-RBS component. We provide a stochastic model to capture the dynamic behaviors to identify the strengths of the promoter-RBS components under the influence of molecular noise. Subsequently, information regarding the strengths of the promoter-RBS components can be constructed as the indexes of promoter-RBS libraries. Here, we describe a standard procedure to characterize these indexes of promoter-RBS components in promoter-RBS libraries.

In this section, we construct three kinds of promoter-RBS libraries according to the characteristic indexes of promoter-RBS components, namely, constitutive, repressor-regulated and activator-regulated promoter-RBS libraries. The anhydrotetracycline (ATc)-responsive library Lib_{TetR} and the isopropyl-β-D-thiogalactopyranoside (IPTG)-responsive library Lib_{LacI} are considered in the repressor-regulated promoter-RBS library $Lib_{repressor}$; the 3-oxy-hexanoyl-homoserine-lactone-responsive library Lib_{LuxR} is considered in the activator-regulated promoter-RBS library $Lib_{activator}$. Simple representative schemes for characterizing the three kinds of promoter-RBS components in the three promoter-RBS libraries are shown in Figure 5.2.

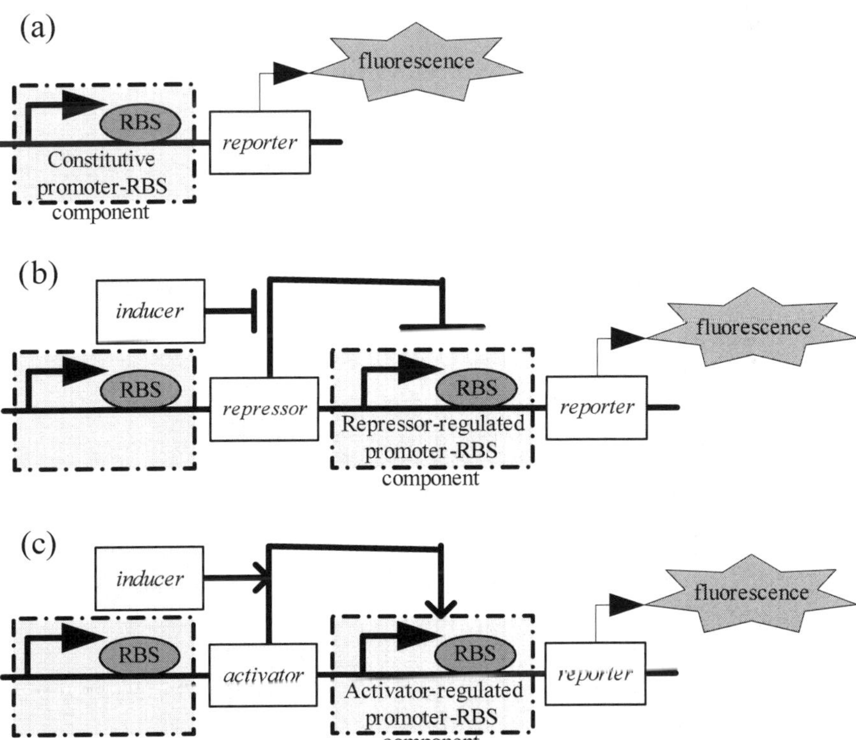

Figure 5.2. Three simple representative schemes for characterizing (a) constitutive, (b) repressor-regulated, and (c) activator-regulated promoter-RBS components. The reporter gene placed downstream of the promoter-RBS component was used to measure the fluorescence of various promoter-RBS components.

In general, transcription of a gene gives rise to mRNA, which is subsequently translated into a protein. The synthesis of mRNA and protein is counter balanced by dilution and degradation of the gene products. These processes determine the net accumulation of mRNA and protein in the cell. The reporter protein is usually used to produce fluorescence which can be measured by ELISA reader. The expression of the reporter gene roughly follows the same stages as those of the protein; however, the maturation time may be different. The response of the reporter protein to light excitation is dependent on post-translational modifications, notably, folding of the protein to the fluorescent form (Tsien 1998). This maturation process gives rise to an additional reaction step from reporter protein to fluorescence, hence the difference between protein and reporter protein (Figure 5.3).

Here, the effect of transcription and translation are integrated as a combined reaction because the half-lives of mRNA are shorter than the corresponding half-lives of proteins. Let us consider the protein that is produced by the promoter-RBS component c, degraded by protease, and

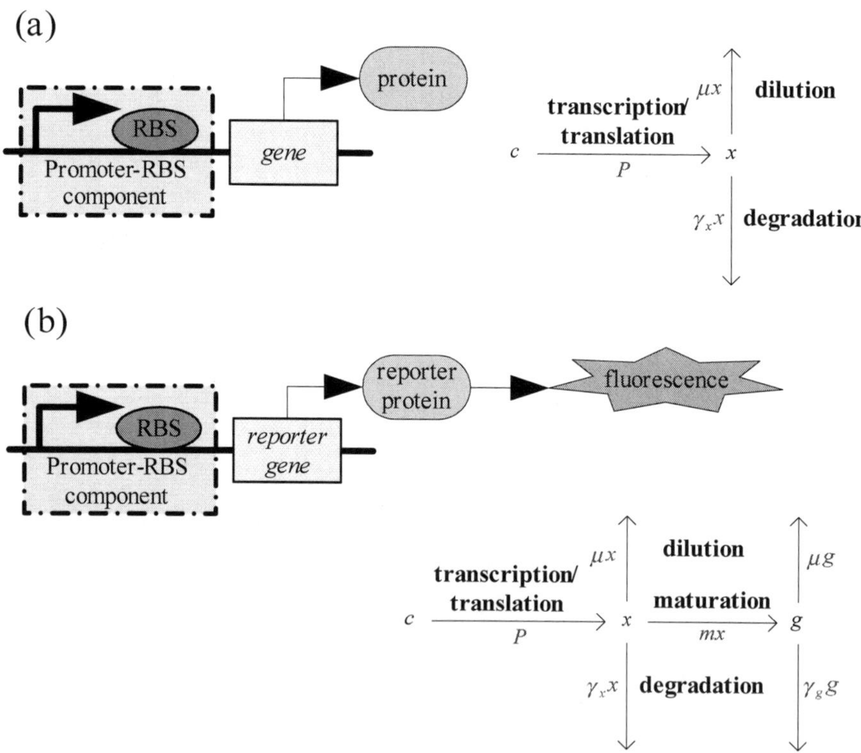

Figure 5.3. Schematic representations of the expressions of (a) protein and (b) reporter protein.

diluted by cell growth. Denoting the concentration of the protein as x, the protein expression arising from a promoter-RBS component c in Figure 5.3a can be described as

$$\dot{x}(t) = P(S_u, S_l, TF, I) - (\mu + \gamma_x)x(t) \tag{5.6}$$

where $x(t)$ denotes the concentration of protein; $P(S_u, S_l, TF, I)$ denotes the promoter-RBS regulation function with the maximum and minimum promoter-RBS strengths S_u and S_l, respectively, from corresponding libraries; TF denotes the related transcription factor concentration; I denotes the related inducer concentration. The dilution rate due to cell growth and the degradation rate of the corresponding protein are denoted by μ and γ_x denote, respectively. On the other hand, the protein expression of a reporter protein produced from a promoter-RBS component c in Figure 5.3b is described as

$$\dot{x}(t) = P(S_u, S_l, TF, I) - (m + \mu + \gamma_i)x(t)$$
$$\dot{g}(t) = mx(t) - (\mu + \gamma_g)g(t) \tag{5.7}$$

where $x(t)$ and $g(t)$ denote the concentrations of immature and mature reporter protein, respectively; γ_i and γ_g denote the degradation rates of immature and mature reporter protein, respectively; m denotes the maturation rate of the reporter protein. In general, the concentration of mature reporter protein is represented by fluorescence, which can be measured by ELISA reader.

When a promoter-RBS component c is placed upstream of the protein, protein production cannot be measured. However, when the promoter-RBS component c is placed upstream of the reporter protein, we can measure the fluorescence by ELISA reader. Hence, we can observe the different protein expressions when a promoter-RBS component c is placed upstream of the protein or reporter protein, which can be modeled by (5.6) and (5.7), respectively. Therefore, from the viewpoint of systems biology, the dynamic characterization of the promoter-RBS component is important and feasible. We can measure fluorescence by ELISA reader to characterize the promoter-RBS component, and subsequently obtain the promoter-RBS strength P of a promoter-RBS component c from the fluorescent data and dynamic model in (5.7). Thereafter, the characteristic indexes of promoter-RBS libraries can be constructed by collecting the promoter-RBS strengths of promoter-RBS components. Note that the strength of the promoter-RBS component c for regulating the protein or reporter protein is the same. Hence, we can also estimate the regulation of the protein if we can identify the promoter-RBS strength from the fluorescent data measured using the reporter protein. According to this assumption, engineering of a synthetic gene circuit with desired behaviors becomes possible if we can select promoter-RBS components with adequate promoter-RBS strengths.

5.2.1 Construction of the constitutive promoter-RBS library

Before identifying the promoter-RBS strength by measuring fluorescence, two important factors should be considered. First, newly synthesized fluorescent protein must undergo a series of self-modifications in order to become fluorescent (Tsien 1998). The appearance of fluorescence reportedly lags about 3.5 hours behind the actual synthesis of the protein (using wild-type green fluorescent protein (GFP) from the jellyfish *Aequorea victoria*) (Heim et al. 1995, Albano et al. 1996). Second, the growth rate of the cell is an important factor. The faster that cells divide, the faster is the dilution of GFP (Leveau and Lindow 2001). Therefore, growth rate should be also considered as a parameter in the stochastic model for promoter-RBS strength identification. The dilution rate μ due to cell growth should be considered initially. Since the cells grow in the same media and condition in the experiments, the dilution rate μ can be considered as the same value. Synthetic biologists can estimate the dilution rate μ in their experiments using different media and experimental conditions. The dilution rate of a cell can be estimated from the cell density by means of the classical formula:

$$\mu = \frac{d \ln s}{dt} = \frac{ds}{dt} \frac{1}{s} \tag{5.8}$$

where s denotes the O.D. 600 value in the experiment. Here, according to the experimental data, the dilution rate is 0.011946 ± 0.000198 min^{-1} at 95% confidence.

On the basis of the above two factors, that is, maturation and dilution rates, and the resistance of the constitutive promoter-RBS component C_i to changes in TF, its downstream gene can be expressed continually at the maximum expression of the component. The stochastic model for protein expression from a constitutive promoter-RBS component (Leveau and Lindow 2001, Wang et al. 2008, Kelly et al. 2009, de Jong et al. 2010) is described as

$$\dot{x}(t) = P_c(S_u, 0, 0, 0) - (m + \mu + \gamma_i)x(t) + \omega_1(t)$$
$$\dot{g}(t) = mx(t) - (\mu + \gamma_g)g(t) + \omega_2(t) \tag{5.9}$$

where $x(t)$ and $g(t)$ denote the concentrations of immature and mature reporter protein, respectively; γ_i and γ_g are the degradation rates of immature and mature reporter protein, respectively; P_c denotes the promoter-RBS strength; m denotes the maturation rate of reporter protein; $\omega_1(t)$ and $\omega_2(t)$ denote environmental noises in the immature and mature reporter proteins, respectively. Further, the promoter-RBS regulation function of constitutive promoter-RBS components C_i from the constitutive promoter-RBS library Lib_{const} can be represented as

$$P_c(S_{u,i},0,0,0) = S_{u,i} \tag{5.10}$$

$$\{S_{u,i}\} \in Lib_{const}$$

where $S_{u,i}$ is the maximum strength of C_i.

Realistically, the mRNA transcripts of the protein, which initiate production of non-fluorescent reporter protein $x(t)$, subsequently mature into fluorescence $g(t)$ that is measured by ELISA reader. In general, GFP is typically used as the reporter protein. According to the literature, the maturation rate m can be calculated as ln2 divided by the time constant of GFP maturation, which has been determined as 2.0 h for wild-type GFP and 0.45 h for faster-folding S65T mutants such as EGFP (Heim et al. 1995). The model in (5.9) is considered to be effective while transcription and translation are integrated as a combined reaction because mRNA half-lives are shorter than the corresponding protein half-lives (typically, the half-life of mRNA of *E. coli* is 6.8 min (Selinger et al. 2003)). This consideration is similar to the results of previous studies (Alper et al. 2005), despite the fact that there are a variety of both promoter and RBS strengths that could be combined to allow synthetic biologists to achieve a range of transcription and translation strengths.

The detailed procedure used to constructed the constitutive promoter-RBS library Lib_{const} is shown in Figure 5.4. The procedure is divided into four steps: (I) selecting of a promoter and a RBS to construct a promoter-

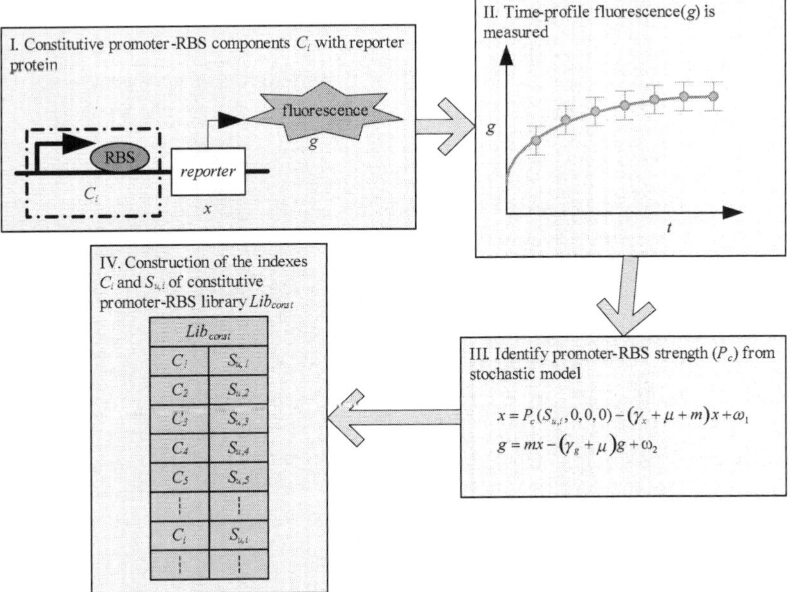

Figure 5.4. The procedure for construction of the constitutive promoter-RBS library Lib_{const}.

RBS component C_i, and then placing them upstream of the reporter protein; (II) measuring the time-profile fluorescence of the selected promoter-RBS component C_i; (III) estimating the dilution rate μ by (5.8), obtaining the remaining parameters m, γ_i and γ_g from the literature (Heim et al. 1995, Leveau and Lindow 2001, Canton et al. 2008, Kelly et al. 2009), and then estimating the promoter-RBS strength $S_{u,i}$ from (5.9) by using identification techniques; (IV) constructing the indexes of the constitutive promoter-RBS library Lib_{const}, including the promoter-RBS component C_i and the identified promoter-RBS strength $S_{u,i}$.

We have listed the constitutive promoter family in Table 5.3. Here, we select four promoters, namely, BBa_J23101, BBa_J23105, BBa_J23106, and BBa_J23114 and eight RBSs, namely, BBa_B0030, BBa_B0032, BBa_B0034, BBa_B0034AA, BBa_B0034TA, BBa_B0034GA, BBa_B0034GC, and BBa_B0034GG to construct 27 promoter-RBS components. Afterward, the promoter-RBS strengths for each promoter-RBS component can be identified by the least-squares method (Johansson 1993). Hence the constitutive promoter-RBS library Lib_{const} is constructed in which the constitutive promoter-RBS components and their strengths are listed (Table 5.4).

5.2.2 Construction of the repressor-regulated promoter-RBS library

The repressor-regulated promoter-RBS component produces the maximum and minimum gene expressions with and without saturation concentrations of the inducer, respectively. Therefore, the concentration of inducer should be considered in the characterization of the repressor-regulated promoter-RBS component, a simple representative scheme for which is shown in Figure 5.2b. As shown in the left of Figure 5.2b, the repressor is produced by the constitutive promoter-RBS component, which can be selected from the constitutive promoter-RBS library Lib_{const}, and then the inducer is added to regulate the concentration of repressor. Finally, the repressor, regulated by the inducer, binds to the repressor-regulated promoter-RBS component R_i to generate fluorescence.

The relationship between the concentrations of the repressor and inducer, which characterizes the repressor-regulated promoter-RBS component, is difficult to determine, and requires data on the relationship between the concentrations of inducer and fluorescence in addition to time-profile data. When the synthetic gene circuit reaches the steady state, different concentrations of inducer produce different fluorescence patterns of reporter protein. Hence, the dose-dependent measurement of inducer and fluorescence is also important for estimating the strength of promoter-RBS components when constructing promoter-RBS libraries.

Table 5.3. List of candidate constitutive promoters and RBSs in BioBrick parts. The promoters and RBSs construct the promoter-RBS components in this section. All BioBrick parts refer to those listed in the Registry of Standard Biological Parts (http://partsregistry.org/Main_Page).

Property	BioBrick part	Description
Constitutive promoter	BBa_J23100	Parts J23100 through J23119 are a family of constitutive promoter parts isolated from a small combinatorial library.
	BBa_J23101	
	BBa_J23102	
	BBa_J23103	
	BBa_J23104	
	BBa_J23105	
	BBa_J23106	
	BBa_J23107	
	BBa_J23108	
	BBa_J23109	
	BBa_J23110	
	BBa_J23111	
	BBa_J23112	
	BBa_J23113	
	BBa_J23114	
	BBa_J23115	
	BBa_J23116	
	BBa_J23117	
	BBa_J23118	
Ribosome binding site (RBS)	BBa_B0030	Parts B0030 through B0034 are all RBSs with different strengths. B0034AA, B0034TA, B0034GA, B0034GC, and B0034GG are variations from B0034.
	BBa_B0031	
	BBa_B0032	
	BBa_B0033	
	BBa_B0034	
	BBa_B0034AA	
	BBa_B0034TA	
	BBa_B0034GA	
	BBa_B0034GC	
	BBa_B0034GG	

For the construction of repressor-regulated promoter-RBS libraries $Lib_{repressor}$, we summarize the construction procedure of $Lib_{repressor}$ as shown in Figure 5.5. Suppose the repressor-regulated promoter-RBS component R_i has maximum and minimum promoter-RBS strengths $S_{u,i}$ and $S_{l,i}$, respectively. The construction of repressor-regulated promoter-RBS libraries $Lib_{repressor}$ can be subsequently divided into two kinds of gene circuit topology. First, the repressor-regulated promoter-RBS component R_i can be constructed following the representative scheme in Figure 5.2a, where

Table 5.4. The constitutive promoter-RBS library Lib_{const}. The promoter-RBS strengths of constitutive promoter-RBS components from L_1 to L_{27} are identified as indexes of the promoter-RBS library Lib_{const} at 95% confidence from the experimental data.

Index	BioBrick component	S_u
C_1	J23101-B0030	1401.1±393.8
C_2	J23101-B0032	156.4±51.3
C_3	J23101-B0034	1286.0±49.1
C_4	J23101-B0034AA	1449.2±33.0
C_5	J23101-B0034TA	357.0±69.0
C_6	J23101-B0034GA	1381.4±49.5
C_7	J23101-B0034GC	854.1±111.2
C_8	J23101-B0034GG	713.6±36.7
C_9	J23105-B0030	226.3±150.7
C_{10}	J23105-B0032	13.6±6.0
C_{11}	J23105-B0034	136.0±6.2
C_{12}	J23105-B0034AA	115.2±9.5
C_{13}	J23105-B0034TA	36.0±3.6
C_{14}	J23105-B0034GA	127.1±6.9
C_{15}	J23105-B0034GC	90.9±5.0
C_{16}	J23105-B0034GG	75.4±3.3
C_{17}	J23106-B0030	366.1±110.3
C_{18}	J23106-B0032	32.6±5.3
C_{19}	J23106-B0034	861.4±56.2
C_{20}	J23106-B0034AA	819.8±22.2
C_{21}	J23106-B0034TA	192.7±23.9
C_{22}	J23106-B0034GA	279.6±13.9
C_{23}	J23106-B0034GC	480.5±26.1
C_{24}	J23106-B0034GG	392.7±46.0
C_{25}	J23114-B0030	46.2±9.5
C_{26}	J23114-B0032	13.3±0.0
C_{27}	J23114-B0034	30.7±3.4

the constitutive promoter-RBS component is replaced by the repressor-regulated promoter-RBS component. Therefore, the maximum promoter-RBS strength $S_{u,i}$ of R_i can be identified using the identification technique in (5.9). In fact, each repressor-regulated promoter-RBS component produces the maximum gene expression with the inducer at saturation concentration. The maximum gene expression should be the same as that produced with

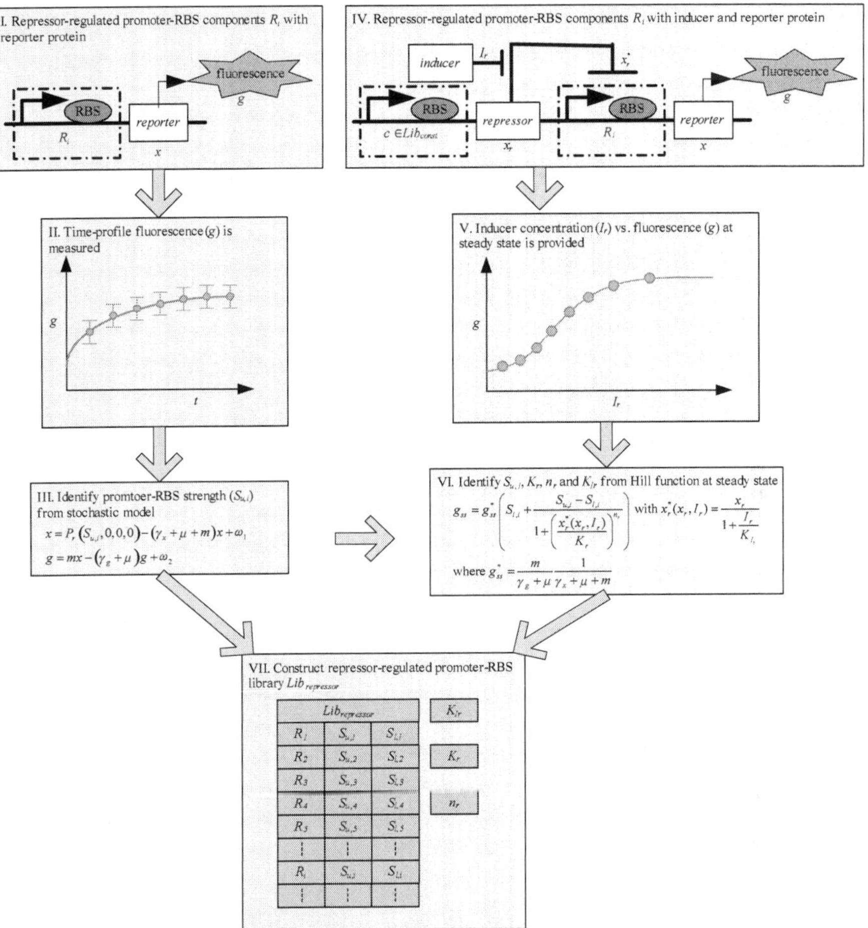

Figure 5.5. The procedure for construction of the repressor-regulated promoter-RBS library $Lib_{repressor}$.

the promoter-RBS strength $S_{u,i}$. Second, according to the representative scheme in Figure 5.2b or Figure 5.5(IV), different concentrations of inducer can drive different fluorescence patterns, implying that different strengths of repressor-regulated promoter RBS components correspond to different concentrations of inducer. In this case, the constitutive promoter selected from the constitutive promoter-RBS library Lib_{const} is used to produce the stable and constitutive repressor x_r, which is regulated by the inducer I_r. In a repressor-regulated promoter-RBS component, x_r represses the gene expression under the control of a repressor-regulated promoter. When the inducer I_r is added, it binds to the repressor x_r and then gene expression under the control of the repressor-regulated promoter occurs. When I_r is removed, x_r becomes active, thereby repressing gene expression again. The repressor activity $x_r^*(x_r, I_r)$ can be described as follows:

$$x_r^*(x_r, I_r) = \frac{x_r}{1 + \dfrac{I_r}{K_{I_r}}} \tag{5.11}$$

where x_r denotes the total repressor concentration, including free and inducer-bound repressor; I_r denotes the inducer concentration for regulating the repressor activity $x_r^*(x_r, I_r)$; K_{I_r} denotes the inducer-repressor dissociation rate.

The promoter-RBS regulation function of repressor-regulated promoter-RBS component R_i dependent on the repressor activity $x_r^*(x_r, I_r)$ can then be represented as follows:

$$P_r(S_{u,i}, S_{l,i}, x_r, I_r) = S_{l,i} + \frac{S_{u,i} - S_{l,i}}{1 + \left(\dfrac{x_r^*(x_r, I_r)}{K_r} \right)^{n_r}} \tag{5.12}$$

$$\{S_{u,i}, S_{l,i}\} \in Lib_{repressor}$$

where K_r and n_r denote the repressor-promoter binding affinity and binding cooperativity between the regulatory protein and repressor-regulated promoter-RBS component, respectively; $S_{u,i}$ and $S_{l,i}$ denote the maximum and minimum repressor-regulated promoter-RBS strengths from the corresponding promoter-RBS library $Lib_{repressor}$, respectively. In this manner, the characteristics of repressor-regulated promoter-RBS components, including the maximum and minimum promoter-RBS strengths as well as the binding affinities, are collected as characteristic indexes of promoter-RBS components to construct the repressor-regulated promoter-RBS libraries, $Lib_{repressor}$.

Here, two repressor-regulated promoter-RBS libraries, Lib_{TetR} and Lib_{LacI}, are constructed by the characteristic indexes of repressor-regulated promoter-RBS components that contain the promoter-RBS strengths of the ATc-responsive promoter P_{tet} and the IPTG-responsive promoter P_{lac} combined with different RBSs. The identification results of the promoter-RBS libraries Lib_{TetR} and Lib_{LacI} are shown in Table 5.5 and Table 5.6, respectively. In the future, synthetic biologists can construct more repressor-regulated promoter-RBS libraries following procedures similar to those described here.

Table 5.5. The repressor-regulated promoter-RBS library Lib_{TetR}. The promoter-RBS strengths of repressor-regulated promoter-RBS components from R_{T1} to R_{T8} are identified as indexes of the promoter-RBS library Lib_{TetR} at 95% confidence from the experimental data.

Index	BioBrick component	S_u	S_l	Parameters
R_{T1}	R0040-B0030	2323.1±259.5	25.5±0.4	$K_{ATc}=0.04337$nM
R_{T2}	R0040-B0032	517.5±17.0	5.6±0.2	$K_{TetR}=148.8$mM
R_{T3}	R0040-B0034	4041.5±247.8	39.6±8.4	$n_{TetR}=1.62$
R_{T4}	R0040-B0034AA	3435.3±210.6	33.66±7.14	
R_{T5}	R0040-B0034TA	1131.6±69.38	11.088±2.35	
R_{T6}	R0040-B0034GA	3475.7±213.1	34.056±7.22	
R_{T7}	R0040-B0034GC	2667.4±163.5	26.136±5.54	
R_{T8}	R0040-B0034GG	3758.6±230.5	36.828±7.81	

Table 5.6. The repressor-regulated promoter-RBS library Lib_{LacI}. The promoter-RBS strengths of repressor-regulated promoter-RBS components from R_{L1} to R_{L8} are identified as indexes of the promoter-RBS library Lib_{LacI} at 95% confidence from the experimental data.

Index	BioBrick component	S_u	S_l	Parameters
R_{L1}	R0010-B0030	9456±487	11.347±2.734	$K_{IPTG}=0.054$nM
R_{L2}	R0010-B0032	195.5±17.7	29.071±15.640	$K_{LacI}=0.4222$mM
R_{L3}	R0010-B0034	7648±152	1.126±0.019	$n_{LacI}=1.363$
R_{L4}	R0010-B0034AA	6531.4±129.2	0.9571±0.0162	
R_{L5}	R0010-B0034TA	2151.5±52.56	0.3153±0.0053	
R_{L6}	R0010-B0034GA	6608.2±130.72	0.9684±0.0163	
R_{L7}	R0010-B0034GC	5071.4±100.32	0.7432±0.0125	
R_{L8}	R0010-B0034GG	7146.1±141.36	1.0472±0.0177	

5.2.3 Construction of the activator-regulated promoter-RBS library

In general, the promoters that can be activated are also used to engineer synthetic gene circuits. By the same manner, an activated promoter, regulated by the inducer, always has the maximum and minimum expression with and without saturation concentrations of the inducers, respectively. Hence, for the construction of activator-regulated promoter-RBS libraries, the concentration of inducer should also be considered when characterizing the activator-regulated promoter-RBS component. A simple representative scheme for characterizing the activator-regulated promoter-RBS component is shown in Figure 5.2c. As shown in the left of Figure 5.2c, the activator is produced by the constitutive promoter-RBS component, which can be selected from the constitutive promoter-RBS library Lib_{const}. The inducer is added to form the protein complex with the activator, and it then binds to the activator-regulated promoter-RBS component.

It is worthy to note that the construction of activator-regulated promoter-RBS libraries differs from the construction of repressor-regulated promoter-RBS libraries. The key point when measuring activator-regulated promoter-RBS components is that there is no fluorescence production when no inducer is added. Hence, the measurement method should be modified to aid construction of the activator-regulated promoter-RBS libraries, the procedure of which is shown in Figure 5.6. Suppose the activator-regulated promoter-RBS component A_i has the maximum and minimum promoter-RBS strengths $S_{u,i}$ and $S_{l,i}$ with and without saturated inducer-bound activator binding, respectively. Therefore, the procedure for constructing the activator-regulated promoter-RBS library $Lib_{activator}$ can be divided into two steps. First, we take specific concentrations of inducers to measure fluorescence and obtain time-profile data of fluorescence. Second, different concentrations of inducer are used to drive different fluorescence patterns, showing different strengths of activator-regulated promoter-RBS components corresponding to different concentrations of the inducer. In this case, the constitutive promoter selected from the constitutive promoter-RBS library Lib_{const} is used to produce the stable and constitutive activator x_a, which is regulated by the inducer I_a. In an activator-regulated promoter-RBS component, x_a activates gene expression under the control of an activator-regulated promoter. The concentration of inducer-bound activator, which is known as the activator activity $x_a^*(x_a, I_a)$ and regulated by external inducer I_a through an inducer binding reaction, can be described as follows:

$$x_a^*(x_a, I_a) = \frac{x_a \cdot I_a}{I_a + K_{I_a}} \tag{5.13}$$

where x_a denotes the total activator concentration including free and inducer-bound activator; I_a denotes inducer concentration for regulating the activator activity $x_a^*(x_a, I_a)$; K_{I_a} denotes the inducer-activator dissociation rate. Therefore, the regulatory strength of activator-regulated promoter-RBS component A_i dependent on the activator activity $x_a^*(x_a, I_a)$ can be represented by a promoter-RBS regulation function as follows (Alon 2007):

$$P_a(S_{u,i}, S_{l,i}, x_a, I_a) = S_{l,i} + \frac{(S_{u,i} - S_{l,i}) \cdot \left(x_a^*(x_a, I_a)\right)^{n_a}}{\left(x_a^*(x_a, I_a)\right)^{n_a} + K_a^{n_a}}$$

$$\{S_{u,i}, S_{l,i}\} \in Lib_{activator} \tag{5.14}$$

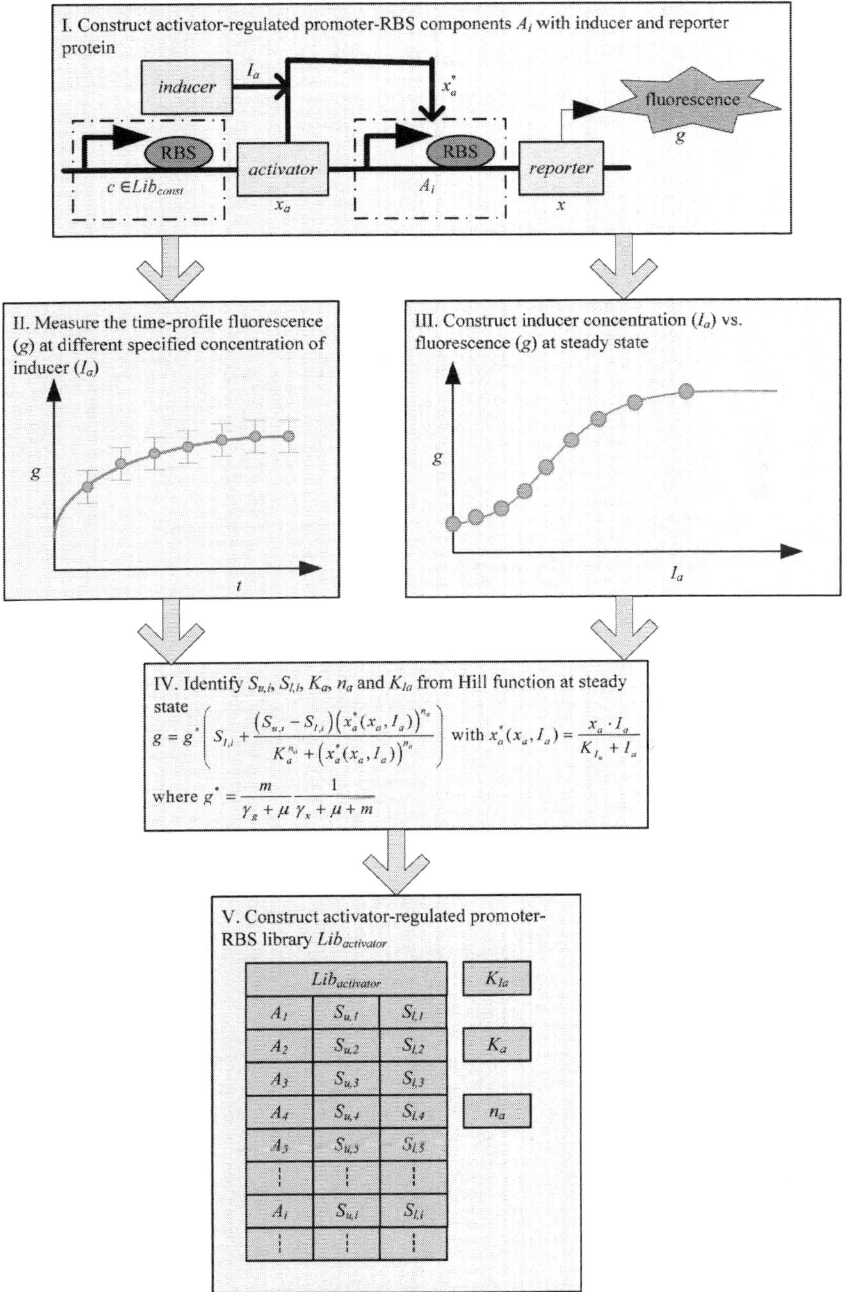

$Lib_{activator}$			K_{Ia}
A_1	$S_{u,1}$	$S_{l,1}$	
A_2	$S_{u,2}$	$S_{l,2}$	K_a
A_3	$S_{u,3}$	$S_{l,3}$	
A_4	$S_{u,4}$	$S_{l,4}$	n_a
A_5	$S_{u,5}$	$S_{l,5}$	
⋮	⋮	⋮	
A_i	$S_{u,i}$	$S_{l,i}$	
⋮	⋮	⋮	

Figure 5.6. The procedure for construction of the activator-regulated promoter-RBS library $Lib_{activator}$.

where K_a and n_a denote the activator-promoter binding affinity and binding cooperativity between the regulatory protein and activator-regulated promoter-RBS component, respectively; $S_{u,i}$ and $S_{l,i}$ denote the maximum and minimum activator-regulated promoter-RBS strengths from the corresponding promoter-RBS library $Lib_{activator}$, respectively. In this manner, the characteristics of activator-regulated promoter-RBS components including the maximum and minimum promoter-RBS strengths and binding affinity are collected to construct the indexes of activator-regulated promoter-RBS libraries.

Here, the promoter-RBS library Lib_{LuxR} of activator-regulated promoter-RBS components, in which the index of the library contains the promoter-RBS strengths of the N-acyl homo serine lactone-responsive promoter P_{lux} combined with different RBSs are constructed. The identification results of promoter-RBS library Lib_{LuxR} are shown in Table 5.7.

Table 5.7. The activator-regulated promoter-RBS library Lib_{LuxR}. The promoter-RBS strengths of activator-regulated promoter-RBS components from R_1 to R_3 are identified as indexes of the promoter-RBS library Lib_{LuxR} at 95% confidence from the experimental data.

Index	BioBrick component	S_u	S_l	Parameters
A_{L1}	R0062-B0030	122100±8400	1160±444.78	K_{AHL}=0.004pM
A_{L2}	R0062-B0032	51430±1610	648±62.37	K_{LuxR}=0.307nM
A_{L3}	R0062-B0034	134900±10300	2010±599.75	n_{LuxR}=1.584

References

Albano, C.R., Randers Eichhorn, L., Chang, Q., Bentley, W.E. and Rao, G. 1996. Quantitative measurement of green fluorescent protein expression. Biotechnology Techniques 10: 953–958.

Alon, U. 2007. An Introduction to Systems Biology: Design Principles of Biological Circuits. Chapman & Hall/CRC, London.

Alper, H., Fischer, C., Nevoigt, E. and Stephanopoulos, G. 2005. Tuning genetic control through promoter engineering. Proceedings of the National Academy of Sciences of the United States of America 102: 12678–12683.

Arkin, A., Ross, J. and McAdams, H.H. 1998. Stochastic kinetic analysis of developmental pathway bifurcation in phage lambda-infected *Escherichia coli* cells. Genetics 149: 1633–1648.

Braun, D., Basu, S. and Weiss, R. 2005. Parameter estimation for two synthetic gene networks: a case study. IEEE International Conference on Acoustics, Speech, and Signal Processing. IEEE.

Canton, B., Labno, A. and Endy, D. 2008. Refinement and standardization of synthetic biological parts and devices. Nature biotechnology 26: 787–794.

Cox, R.S., 3rd, Surette, M.G. and Elowitz, M.B. 2007. Programming gene expression with combinatorial promoters. Mol Syst Biol 3: 145.

de Jong, H., Ranquet, C., Ropers, D., Pinel, C. and Geiselmann, J. 2010. Experimental and computational validation of models of fluorescent and luminescent reporter genes in bacteria. BMC Syst Biol 4: 55.

Ellis, T., Wang, X. and Collins, J.J. 2009. Diversity-based, model-guided construction of synthetic gene networks with predicted functions. Nat Biotechnol 27: 465–471.

Hammer, K., Mijakovic, I. and Jensen, P.R. 2006. Synthetic promoter libraries—tuning of gene expression. Trends Biotechnol 24: 53-55.

Heim, R., Cubitt, A.B. and Tsien, R.Y. 1995. Improved green fluorescence. Nature 373: 663–664.

Iadevaia, S. and Mantzaris, N.V. 2006. Genetic network driven control of PHBV copolymer composition. Journal of biotechnology 122: 99–121.

Johansson, R. 1993. System modeling & identification. Prentice-Hall International, New Jersey.

Kelly, J.R., Rubin, A.J., Davis, J.H., Ajo-Franklin, C.M., Cumbers, J., Czar, M.J., de Mora, K., Glieberman, A.L., Monie, D.D. and Endy, D. 2009. Measuring the activity of BioBrick promoters using an *in vivo* reference standard. J Biol Eng 3: 4.

Leveau, J.H. and Lindow, S.E. 2001. Predictive and interpretive simulation of green fluorescent protein expression in reporter bacteria. J Bacteriol 183: 6752–6762.

Murphy, K.F., Balazsi, G. and Collins, J.J. 2007. Combinatorial promoter design for engineering noisy gene expression. Proc Natl Acad Sci USA 104: 12726–12731.

Rosenfeld, N., Young, J.W., Alon, U., Swain, P.S. and Elowitz, M.B. 2007. Accurate prediction of gene feedback circuit behavior from component properties. Mol Syst Biol 3: 143.

Selinger, D.W., Saxena, R.M., Cheung, K.J., Church, G.M. and Rosenow, C. 2003. Global RNA half-life analysis in *Escherichia coli* reveals positional patterns of transcript degradation. Genome research 13: 216–223.

Tsien, R.Y. 1998. The green fluorescent protein. Annual Review of Biochemistry 67: 509–544.

Tuttle, L.M., Salis, H., Tomshine, J. and Kaznessis, Y.N. 2005. Model-Driven Designs of an Oscillating Gene Network. Biophysical Journal 89: 3873–3883.

Wang, X., Errede, B. and Elston, T.C. 2008. Mathematical analysis and quantification of fluorescent proteins as transcriptional reporters. Biophys J 94: 2017–2026.

Wu, C.H., Lee, H.C. and Chen, B.S. 2011a. Robust synthetic gene network design via library-based search method. Bioinformatics 27: 2700–2706.

Wu, C.H., Zhang, W. and Chen, B.S. 2011b. Multiobjective H2/Hinfinity synthetic gene network design based on promoter libraries. Math Biosci 233: 111–125.

6

Robust Synthetic Gene Network Designs based on Library-search Method

After we have redefined the promoter and promoter-RBS libraries, we propose some library-based search methods to efficiently select adequate promoters/promoter-RBSs from the corresponding libraries to engineer a synthetic gene network with desired behaviors despite intrinsic fluctuations and environmental disturbances on the host cell.

6.1 H_2 Optimal Synthetic Gene Network Design: Promoter Library-based Search Method

6.1.1 H_2 optimal synthetic gene network design and design procedure

First, we consider a simple two-stage transcriptional cascade in Figure 6.1 (Hooshangi et al. 2005). The *yegfp* expression is controlled by LacI protein, which can be repressed by the TetR repressor. TetR is expressed constitutively from the promoter selected from Lib_{const}. When IPTG is added to the growth medium, IPTG binds to the LacI repressor and then induces the *yegfp* expression.

Suppose $x_1(c_1, t)$, $x_2(c_2, t)$, $x_3(c_3, t)$, $X_1(c_1, t)$, $X_2(c_2, t)$, and $X_3(c_3, t)$ are the concentrations of mRNA *tetR*, *lacI*, *yegfp* and proteins TetR, LacI, and yEGFP, respectively. The dynamic model in Figure 6.1 with promoters c_1, c_2 and c_3, selected from the promoter library Lib_{const}, Lib_{TetR}, Lib_{LacI} in Table 5.1, respectively, is modeled as follows (Gardner et al. 2000, Kobayashi et al. 2003)

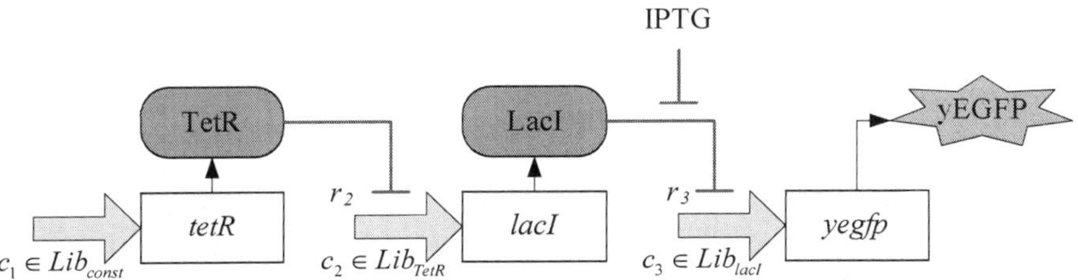

Figure 6.1. A simple synthetic transcriptional cascade. The protein TetR, expressed constitutively from the constitutive promoter c_1, inhibits the production of LacI by binding the TetR-regulated promoter c_2, and LacI inhibits the production of yEGFP by binding the LacI-regulated promoter c_3.

$$\begin{cases} \dot{x}_i\left(c_i,t\right)= p_j\left(c_i,r_i\right)- \beta x_i\left(c_i,t\right) \\ \dot{X}_i\left(c_i,t\right)= \alpha x_i\left(c_i,t\right)- \gamma_i X_i\left(c_i,t\right) \end{cases}$$

$$y\left(c,t\right)= X_3\left(c_3,t\right), i=1,\,2,\,3 \tag{6.1}$$

$$p_j\left(c_i,r_i\right)\in \left\{ p_{const}\left(c_1,0\right), p_{TetR}\left(c_2,X_1\right), p_{LacI}\left(c_3,X_2\right)\right\}$$

$$c = \left\{c_1,c_2,c_3\right\}\in \left\{Lib_{const}, Lib_{TetR}, Lib_{LacI}\right\}$$

where $y(c,t)$ denotes the desired concentration of yEGFP; $\{\gamma_1,\, \gamma_2,\, \gamma_3\} = \{\gamma_{TetR},\, \gamma_{LacI},\, \gamma_{yEGFP}\}$ are protein degradation rates; and the three promoters c_1, c_2 and c_3, are considered as a promoter set, $c = \{c_1,\, c_2,\, c_3\}$.

In general, biological parts or components are inherently uncertain in this network. We assume that the promoter activities, degradation rates of mRNAs and proteins, and translation rates are stochastically uncertain due to gene expression noises in transcriptional and translational processes, thermal fluctuations, DNA mutations and evolutions (Alon 2007, Wu et al. 2011a, Wu et al. 2011b) as follows:

$$c_i = \left\{ c_{r_i},c_{s_i} \right\}\to c_i + \Delta c_i n_i\left(t\right)= \left\{ c_{r_i},c_{s_i} \right\} + \left\{ \Delta c_{r_i}, \Delta c_{s_i} \right\}n_i\left(t\right)$$

$$\beta \to \beta + \Delta \beta n_i\left(t\right),\, \alpha \to \alpha + \Delta \alpha n_i\left(t\right),\, \gamma_i \to \gamma_i + \Delta \gamma_i n_i\left(t\right), i=1,\,2,\,3 \tag{6.2}$$

where $\Delta c_{r_i}, \Delta c_{s_i}, \Delta \beta$, and $\Delta \alpha$ denote the amplitudes of stochastic parameter variations, and $n_i\left(t\right)$ is a Gaussian noise with zero mean and unit variance.

Synthetic gene networks also suffer from environmental disturbances on the host cell, as follows

$$\begin{cases} \dot{x}_i\left(c_i,t\right)= p_j\left(c_i,r_i\right)-\beta x_i\left(c_i,t\right)+\left(p_j\left(\Delta c_i,r_i\right)-\Delta\beta x_i\left(c_i,t\right)\right)n_i\left(t\right)+v_i\left(t\right) \\ \dot{X}_i\left(c_i,t\right)= \alpha x_i\left(c_i,t\right)-\gamma_i X_i\left(c_i,t\right)+\left(\Delta\alpha x_i\left(c_i,t\right)-\Delta\gamma_i X_i\left(c_i,t\right)\right)n_i\left(t\right)+v_i\left(t\right) \end{cases}$$

$$y\left(c,t\right)= X_3\left(c_3,t\right), i=1, 2, 3$$

$$p_j\left(c_i,r_i\right)\in \left\{p_{const}\left(c_1,0\right), p_{TetR}\left(c_2,X_1\right), p_{LacI}\left(c_3,X_2\right)\right\}$$

$$c=\left\{c_1,c_2,c_3\right\}\in \left\{Lib_{const}, Lib_{TetR}, Lib_{LacI}\right\}$$

$$(6.3)$$

The design purpose is to select the most adequate promoter set $c^* = \left\{c_1^*,c_2^*,c_3^*\right\}$ from corresponding promoter libraries so that the observed output $y(c^*,t)$ can track the following desired reference trajectory.

$$y_r\left(t\right)= f_r\left(t\right) \tag{6.4}$$

Hence, in order to engineer a stochastic synthetic gene network in (6.3) which can robustly and optimally track a desired trajectory (6.4), a library-based search method is provided to efficiently select the most adequate promoter set $c^* = \left\{c_1^*,c_2^*,c_3^*\right\}$ from promoter libraries.

Consider a more general design for a synthetic gene network. We extend the above gene network in Figure 6.1 with promoters from constitutive, TetR- and LacI-regulated promoter libraries in Table 5.1 to an n-gene network with promoters from m promoter libraries Lib_j, $j=1,2,\ldots m$ based on a prescribed genetic circuit topology. In the n-gene network, the promoter regulation function for the promoter can be represented as follows (Wu et al. 2011a, Wu et al. 2011b)

$$p_j\left(c_i,r_i\right)=c_{r_i} +\frac{c_{s_i}-c_{r_i}}{1+\left(r_i/K_j\right)^{n_j}}$$

$$(6.5)$$

$$c_i = \left\{c_{r_i},c_{s_i}\right\}\in Lib_j, j=1, 2, \ldots, m$$

where K_j and n_j are the corresponding parameters to be identified or estimated. For example, the parameters K_{TetR} and n_{TetR} are for the TetR-regulated promoter library estimated and listed in Table 5.2. Hence, a dynamic model of the n-gene circuit topology with n promoters selected from m promoter libraries with intrinsic parameter fluctuations and environmental disturbances is represented as (Wu et al. 2011a)

$$\dot{X}\left(c,t\right)= f\left(X,c,t\right)+\Delta f\left(X,c,t\right)n\left(t\right)+v\left(t\right)$$

$$y\left(c,t\right)= HX\left(c,t\right)$$

$$(6.6)$$

$$c=\left\{c_1,c_2,\ldots,c_n\right\}\in Lib_j, j=1, 2, \ldots, m$$

where $x_i(c_i,t)$ and $X_i(c_i,t)$ denote the concentrations of mRNA and proteins in the synthetic gene network, respectively. $X(c,t) = [x_1(c,t),X_1(c,t),\ldots,x_n(c,t),X_n(c,t)]^T$ denotes the concentrations of mRNAs and proteins in the synthetic gene network; $f(X,c,t)$ denotes the nonlinear gene regulation of transcription and translation based on the circuit topology; $\Delta f(X,c,t)$ denotes the parameter fluctuations of the nonlinear gene regulation $f(X,c,t)$; and $y(c,t)=[y_1(c,t),\ldots,y_p(c,t)]^T$ denotes the observed proteins we are interested in. H is a $p \times n$ output matrix determining what proteins we are interested in.

The stochastic dynamic system in (6.6) can be established from the prescribed genetic circuit topology, and the most adequate promoter set $c^* = \{c_1^*, c_2^*, \ldots, c_n^*\}$ selected from corresponding promoter libraries Lib_j, $j = 1, 2, \ldots, m$ can make the synthetic gene network track the prescribed desired trajectories via the library-based search method. For convenience in illustrating the proposed library-based search method, four design specifications are given as follows (Wu et al. 2011a):

i) Given the desired reference trajectory $y_r(t) = f_r(t)$, which dimension is consistent with the numbers of fluorescent proteins $y(c,t)$ in (6.6).

ii) Given the promoter libraries Lib_j, $j = 1, \ldots, m$ as in Table 5.1.

iii) Given the environmental disturbances $v_i(t)$ and standard deviation of promoter activities, $\Delta c_i = \{\Delta c_{r_i}, \Delta c_{s_i}\}$, of degradation rates of mRNAs and proteins, $\Delta \beta$ and $\Delta \gamma_i$, and of translation rates, $\Delta \alpha$, to be tolerated in the host cell.

iv) Select a promoter set $c = \{c_1, c_2, \ldots, c_n\}$ from the promoter libraries Lib_j, $j = 1, \ldots, m$ to minimize the following cost function, i.e.,

$$J(c) = E \int_0^{t_f} \left(y(c,t) - y_r(t)\right)^T \left(y(c,t) - y_r(t)\right) dt \tag{6.7}$$

If the cost function in (6.7) can be minimized by selecting the most adequate promoter set $c^* = \{c_1^*, c_2^*, \ldots, c_n^*\}$ under design specifications (i)–(iv), then the observed output $y(c^*,t)$ of the synthetic gene network can optimally and robustly track the prescribed reference trajectories $y_r(t)$ under the intrinsic parameter fluctuations and environmental disturbances on the host cells.

Using conventional optimal design methods, many combinations of promoter sets required to satisfy four design specifications generally will waste a large amount of computation time and trial-and-error experiments. In this section, a library-based search method using a genetic algorithm (GA) is employed to select the most adequate promoter set c^* from corresponding promoter libraries. GA is a stochastic optimization algorithm, originally motivated by the mechanisms of natural selection and evolutionary genetics.

GA has been proven to be efficient for solving constrained optimization problems in many areas (Grefenstette 1986) and further details on GA can be available elsewhere (Goldberg 1989).

By the library-based search method and four design specifications, the most adequate promoter set c^* can be selected in design specification (ii). Despite intrinsic parameter fluctuations and environmental disturbances in design specifications (iii), the gene network can achieve the desired reference trajectory in design specification (i) by minimizing the cost function $J(c)$ in design specification (iv). In fact, this library-based optimal reference tracking problem in (6.7) is a highly nonlinear optimization problem. In conventional search algorithms, the optimal solution can be obtained, but it may not be the global optimization. The library-based search method using GA is an iterative procedure to select the most adequate promoter set $c^* = \{c_1^*, c_2^*, \ldots, c_n^*\}$ that satisfies design specifications (i)–(iv). When the most adequate promoter set c^* is selected, design specifications (i)–(iv) for the synthetic gene network can be satisfied, and then the synthetic gene network can track the desired reference trajectory robustly and optimally.

A design procedure for this library-based search method using GA is given as follows (Wu et al. 2011a).

1) Build the redefined promoter libraries from the experimental data of maximum and minimum outputs and input-output relationship between the fluorescence, repressor and inducer.
2) Construct a genetic circuit topology such as Figure 6.1 with promoters, and then build the stochastic dynamic model in (6.6) for the synthetic gene network.
3) Provide the design specifications (i)–(iv) for the synthetic gene network in (6.6).
4) Generate initial promoter sets from the redefined promoter libraries.
5) Calculate the cost function $J(c)$ in (6.7) for each promoter set c in the population.
6) Create offspring by GA operator (i.e., reproduction, crossover and mutation).
7) Evaluate the new promoter sets and calculate the cost function of each promoter set obtained by natural selection.
8) Stop if the search goal is achieved, or an allowable generation is attained; else go to step 6.

In each iteration or generation of GA, these genetic operators are performed to generate new populations (i.e., promoter sets), and these new populations are evaluated via the cost function in (6.7). On the basis of these genetic operators and evaluations, a better new population of candidate solutions is formed in each genetic generation.

6.1.2 In silico design examples

In this section, two *in silico* design examples are given to illustrate the design procedure of the proposed library-based search method. First of all, synthetic biologists need to engineer a genetic circuit topology, and then a dynamic model could be constructed. Finally, the proposed library-based search method is used to obtain the most adequate promoter set $c^* = \{c_1^*, c_2^*, \ldots, c_n^*\}$.

6.1.2.1 Design of synthetic transcription cascade

Consider the dynamic system (6.3) of the synthetic transcriptional cascade shown in Figure 6.1. The design specifications are provided and listed as follows:

i) The desired reference trajectory to be tracked is

$$y_r(t) = \begin{cases} 10^6, & t \le 40, t > 80 \\ 2 \times 10^7, & 40 < t \le 80 \end{cases} \tag{6.8}$$

   i.e., the reference trajectory $y_r(t)$ is high from 40 to 80 hours and low at other times. In this design example, 10 mM IPTG is added to induce the network transition from 40 hours and is removed at 80 hours. The concentration of yEGFP will track the reference trajectory $y_r(t)$ in (6.8).

ii) Since $c_1 \in Lib_{const}$, $c_2 \in Lib_{TetR}$ and $c_3 \in Lib_{LacI}$, $c_1 \in J_m$, $c_2 \in T_k$ and $c_3 \in L_k$, $m = 0, 1, \ldots, 18$, $k = 0, 1, \ldots, 20$ are looked up in Table 5.1.

iii) Suppose the standard deviations of parameter fluctuations $\Delta c_i = \{\Delta c_{r_i}, \Delta c_{s_i}\} = \{0.1 c_{r_i}, 0.1 c_{s_i}\}$, $i = 1, 2, 3$, $\Delta d = 0.1 \times d$, $\Delta \alpha = 0.1 \times \alpha$, $\Delta \gamma_1 = 0.1 \times \gamma_{TetR}$, $\Delta \gamma_2 = 0.1 \times \gamma_{LacI}$, $\Delta \gamma_3 = 0.1 \times \gamma_{yEGFP}$ and environmental disturbances $v_i(t)$ are independent Gaussian white noises with zero means and unit variances.

iv) The following mean-square tracking error needs to be minimized

$$J(c) = \min_{c = \{c_1, c_2, c_3\}} E \int_0^{t_f} \left(y(c, t) - y_r(t) \right)^2 dt \tag{6.9}$$

In order to solve the constrained optimal tracking design problem of synthetic gene network via the proposed library-based search method, GA operators are set as follows: (1) a roulette wheel selection is used to increase the selection efficiency of the population with a lower cost function score; (2) the crossover rate is 0.8; (3) the chromosome mutates uniformly with the mutation rate 0.1. Then the most adequate promoter set

$c^* = \{c_1^*, c_2^*, c_3^*\} = \{J_4, T_2, L_1\}$ can be obtained. The simulation result shows that the proposed library-based search method can efficiently find the most adequate promoter set c^* (see Figure 6.2).

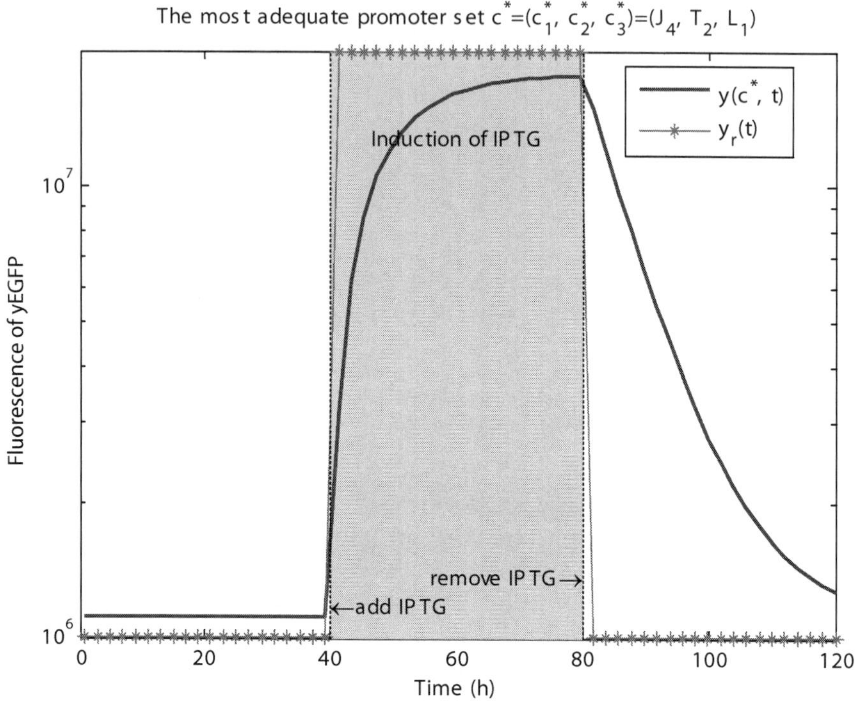

Figure 6.2. Simulation results for synthetic transcription cascade design. The most adequate promoter set $c^* = \{c_1^*, c_2^*, c_3^*\} = \{J_4, T_2, L_1\}$ is obtained via library-based search method through GA. The system behavior $y(c^*, t)$ of the gene network employs the most adequate promoter set c^* to track the desired reference trajectory $y_r(t)$, which is at high from 40 to 80 hours and at low in other hours. IPTG is added from 40 hours and removed from 80 hours.

Color image of this figure appears in the color plate section at the end of the book.

6.1.2.2 Design of synthetic genetic oscillator

Consider a synthetic gene oscillator with negative feedback loops shown in Figure 6.3 (Elowitz and Leibler 2000). The repressor protein LacI inhibits the expression of the gene *cI*, whose protein product in turn inhibits the expression of the gene *tetR*. The repressor protein TetR inhibits the transcription of *lacI* and *yegfp*. The negative feedback loops lead to temporal oscillations if the adequate promoter set is selected. For convenience, the concentrations of mRNA: *lacI*, *cI*, *tetR*, *yegfp* and proteins: LacI, CI, TetR,

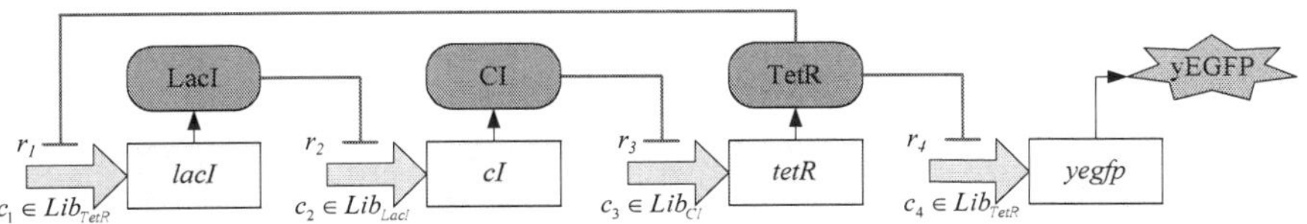

Figure 6.3. The synthetic gene oscillator. The synthetic oscillator constructed by four repressor-regulated promoters c_1, c_2, c_3 and c_4, where $c_1, c_4 \in T_k$ and $c_2 \in L_k$, $k = 0, 1, \ldots, 20$ in Table 5.1. $c_3 \in Lib_{CI}$ is a CI-regulated promoter that has no library at present, so we suppose the promoter regulation function has a fixed from as (6.11).

yEGFP are denoted by $x_1(c_1,t)$, $x_2(c_2,t)$, $x_3(c_3,t)$, $x_4(c_4,t)$, and $X_1(c_1,t)$, $X_2(c_2,t)$, $X_3(c_3,t)$, $X_4(c_4,t)$, respectively. The dynamic system for circuit topology in Figure 6.3 is given as

$$
\begin{cases}
\dot{x}_i\left(c_i,t\right)= p_j\left(c_i,r_i\right)- \beta x_i\left(c_i,t\right) \\
\dot{X}_i\left(c_i,t\right)= \alpha x_i\left(c_i,t\right)- \gamma_i X_i\left(c_i,t\right)
\end{cases}
$$

$$
y\left(c,t\right)= X_4\left(c_4,t\right), i = 1, 2, 3, 4 \tag{6.10}
$$

$$
p_j\left(c_i,r_i\right)\in \left\{p_{TetR}\left(c_1,X_3\right), p_{LacI}\left(c_2,X_1\right), p_{CI}\left(c_3,X_2\right), p_{TetR}\left(c_4,X_3\right)\right\}
$$

$$
c = \left\{c_1,c_2,c_3\right\}\in \left\{Lib_{TetR}, Lib_{LacI}, Lib_{TetR}\right\}
$$

Since there is no CI-regulated promoter library at present, we suppose that the promoter regulation function for CI has a fixed form as

$$
p_{CI}\left(c_3,X_2\right)=\frac{150}{1+\left(X_2 / K_{CI}\right)^{n_{CI}}} \tag{6.11}
$$

where K_{CI} and n_{CI} are binding affinity and binding cooperativity between CI and DNA, respectively.

Based on the synthetic gene network in (6.10) with intrinsic fluctuations and environmental disturbances, the design specifications are provided and listed as follows:

i) The desired reference trajectory to be tracked is
$$
y_r\left(t\right)= 4000\sin\left(0.016\pi t\right)+ 22000 \tag{6.12}
$$

ii) The indexes of TetR- and LacI-regulated promoter libraries in Table 5.1 are both from 0 to 20. The promoter regulation function for the CI-regulated promoter library is shown in (6.11) with $K_{CI} = 20$ and $n_{CI} = 2$.

iii) According to the values of system parameters, we assume the parameter fluctuations with the standard deviations $\Delta c_i = \{\Delta c_{r_i}, \Delta c_{s_i}\} = \{0.1c_{r_i}, 0.1c_{s_i}\}$, $i = 1,\ldots,4$, $\Delta d = 0.1d = 0.0288$, $\Delta \alpha = 0.1\alpha = 2.4$, $\Delta \gamma_1 = 0.1\gamma_{LacI}$, $\Delta \gamma_2 = 0.1\gamma_{cI}$, $\Delta \gamma_3 = 0.1\gamma_{TetR}$ and $\Delta \gamma_4 = 0.1\gamma_{yEGFP}$.

iv) The following mean-square tracking error needs to be minimized

$$J(c) = \min_{c=\{c_1,c_2,c_4\}} E \int_0^{t_f} \left(y(c,t) - y_r(t) \right)^2 dt \tag{6.13}$$

The proposed library-based search method is used to solve the constrained optimal tracking design problem through GA. GA operators are set as follows: (1) a roulette wheel selection is used to increase the selecting efficiency of the population with a lower cost function score; (2) the crossover rate is 0.8; (3) the chromosome mutates uniformly with the mutation rate 0.1. Then the most adequate promoter set $c^* = \{c_1^*, c_2^*, c_4^*\} = \{T_{17}, L_{19}, T_8\}$ can be obtained (see Figure 6.4).

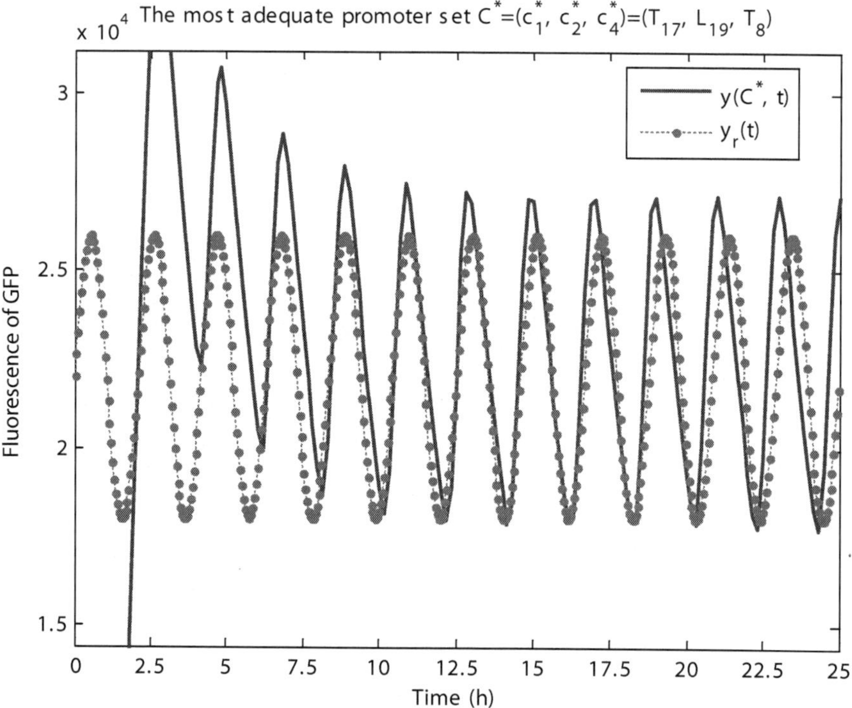

Figure 6.4. Simulation results for synthetic genetic oscillator design. The most adequate promoter set $c^* = \{c_1^*, c_2^*, c_4^*\} = \{T_{17}, L_{19}, T_8\}$ for the synthetic oscillator is selected using the library-based search method. The oscillatory behavior $y(c^*,t)$ with the most adequate promoter set c^* can track the desired reference trajectory $y_r(t)$.

Color image of this figure appears in the color plate section at the end of the book.

6.1.3 Summary

The main challenge in genetic circuit design lies in selecting well-matched genetic parts that combine and produce the desired behavior reliably. Although the parameter values can be calculated by a stochastic dynamic model (Chen and Wu 2009), it is hard to select the biological part that implements a desired cellular function with quantitative values. To overcome this problem, synthetic biologists usually create many versions of synthetic circuits with diverse characteristic by directed evolution, point mutation or random combinational of DNA components, and the functions of these versions are investigated to engineer the gene circuit with the desired behavior. But when the design of the genetic circuit is complex, the number of mutated versions needed to be created and tested is dramatically increased. Hence these experimental steps become tedious and time-consuming due to the significant amount of trail-and-error experiments.

In this section, based on the redefined promoter libraries, a library-based search method is introduced to engineer a synthetic gene network with desired behaviors by satisfying four design specifications. Through re-characteristic biological part data sheets, we can design a synthetic gene network with a new function in the existing libraries without having to perform a large number of trial-and-error experiments. Using GA operators, the library-based search method is used to mimic the realistic behavior in the host cell. Hence the proposed method can efficiently engineer a synthetic gene network to perform its desired behaviors despite intrinsic parameter fluctuations and environmental disturbances.

In design examples of the synthetic gene network, the most adequate promoter set c^* is selected to achieve the minimum tracking error between the desired reference trajectory and the observed output (i.e., fluorescence) despite intrinsic parameter fluctuations and environmental disturbances *in vivo*, as shown in Figure 6.2. It is worth to notice that this method is also useful for constructing a synthetic oscillator (Figure 6.3). After constructing the synthetic gene network and providing four design specifications, the library-based search method can also efficiently select the most adequate promoter set to track the desired trajectory, as shown in Figure 6.4. Based on the proposed method and the design examples, we have demonstrated that the design procedure of a synthetic gene network can be simplified for *in vivo* experiments.

6.2 Multiobjective H_2/H_∞ Reference Tracking Design: Promoter Library-based Search Method

6.2.1 Multiobjective H_2/H_∞ reference tracking design and design procedure

First, for the convenience of illustration, the simple toggle switch in Figure 6.5 is given as a design example. The toggle switch has two distinct stable states and can be reversibly switched between the two states by changing the inducers ATc and IPTG. The proteins TetR and LacI inhibit transcription through the promoters c_2 and c_1 and are induced by the inducers ATc and IPTG, respectively. The fluorescent protein yEGFP is repressed through the promoter c_3 by the repressor TetR. Assume that $x_1(c_1,t)$, $x_2(c_2,t)$ and $x_3(c_3,t)$ denote the concentrations of mRNAs *tetR*, *lacI* and *yegfp*, respectively. In addition, $X_1(c_1,t)$, $X_2(c_2,t)$ and $X_3(c_3,t)$ denote the concentrations of proteins TetR, LacI and yEGFP, respectively. Then the dynamic model of the toggle switch gene network in Figure 6.5 is modeled as follows (Gardner et al. 2000, Kobayashi et al. 2003, Wu et al. 2011b)

$$\begin{cases} \dot{x}_1(c_1,t) = p_{LacI}(c_1,r_1) - \beta x_1(c_1,t) \\ \dot{X}_1(c_1,t) = \alpha x_1(c_1,t) - \gamma_1 X_1(c_1,t) \\ \dot{x}_2(c_2,t) = p_{TetR}(c_2,r_2) - \beta x_2(c_2,t) \\ \dot{X}_2(c_2,t) = \alpha x_2(c_2,t) - \gamma_2 X_2(c_2,t) \\ \dot{x}_3(c_3,t) = p_{TetR}(c_3,r_3) - \beta x_3(c_3,t) \\ \dot{X}_3(c_3,t) = \alpha x_3(c_3,t) - \gamma_3 X_3(c_3,t) \end{cases} \tag{6.14}$$

$$y(c,t) = X_3(c_3,t)$$

$$c = \{c_1, c_2, c_3\}, \; c_1 \in Lib_{LacI}, \; c_2, c_3 \in Lib_{TetR}$$

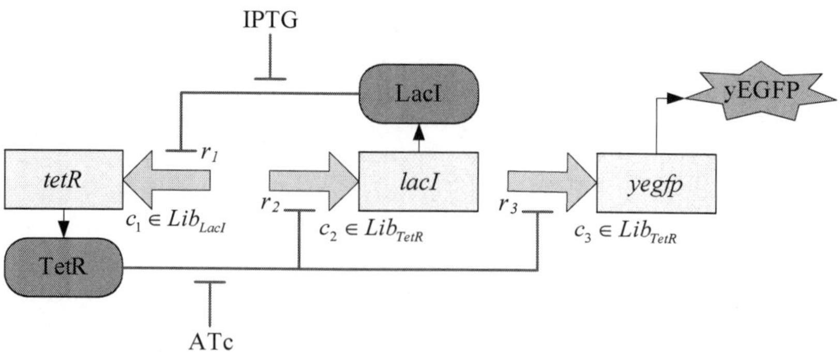

Figure 6.5. Synthetic toggle switch.

with $r_1 = X_2(c_2,t) / (1 + (IPTG / K_{IPTG})^{n_{IPTG}})$ and $r_2 = r_3 = X_1(c_1,t)/(1 + (ATc/ K_{ATc})^{nATc})$ for the gene circuit topology in Figure 6.5, where α denotes the translation rate; and β and γ_i, $i = 1, 2, 3$ denote the degradation rates for mRNA and proteins, respectively. The promoter regulation functions $p_{LacI}(c_1,r_1)$, $p_{TetR}(c_2,r_2)$ and $p_{TetR}(c_3,r_3)$ are dependent on the selection of promoters c_1, c_2 and c_3 from the corresponding promoter libraries. The output $y(c,t)$ denotes the output of interest and is dependent on the selected promoter set $c = \{c_1,c_2,c_3\}$ with adequate promoter activities from the corresponding promoter libraries in Table 5.1. Obviously, the dynamic model of toggle switch gene network (6.14) consists of three interactive dynamic models of promoter-regulation gene network in (5.3).

In general, biological parts are inherently uncertain in this nano-scale biochemical system. We assume that the promoter activities of promoters, degradation rates of mRNAs and proteins, and translation rates are stochastically uncertain *in vivo* due to gene expression noises in transcriptional and translational processes, thermal fluctuations, DNA mutations, parameter estimation errors and evolutions (Alon 2007, Wu et al. 2011a, Wu et al. 2011b) as follows:

$$c_{r_k} \to c_{r_k} + \Delta c_{r_k} n_k(t),\, c_{s_k} \to c_{s_k} + \Delta c_{s_k} n_k(t),\, \gamma_k \to \gamma_k + \Delta \gamma_k n_k(t)$$
$$\beta \to \beta + \Delta\beta n_k(t),\, \alpha \to \alpha + \Delta\alpha n_k(t),\, k = 1, 2, 3 \tag{6.15}$$

where Δc_{r_k}, Δc_{s_k}, $\Delta \gamma_k$, $\Delta\beta$ and $\Delta\alpha$ denote the standard deviations of stochastic parameter, and $n_k(t)$ is a Gaussian noise with zero mean and unit variance. Thus Δc_{r_k}, Δc_{s_k}, $\Delta \gamma_k$, $\Delta\beta$ and $\Delta\alpha$ denote the deterministic parts of parameter variations and $n_k(t)$ denotes the kth random fluctuation source.

Suppose the synthetic gene network also suffers from environmental disturbances on the host cell due to various extracellular environments in the cellular context. Then the synthetic gene network in (6.14) with a promoter set $c = \{c_1,c_2,c_3\}$ selected from the redefined promoter libraries, with intrinsic parameter fluctuations as (6.15) and environmental disturbances on the host cell could be described as (Wu et al. 2011b)

$$\dot{X}(c,t) = f(X,c,r,t) + \sum_{k=1}^{3} M_k g_k(X,c_k,r_k,t)\, n_k(t) + v(t)$$
$$Y(c,t) = HX(c,t) \tag{6.16}$$
$$c = \{c_1,c_2,c_3\},\, c_1 \in Lib_{LacI},\, c_2,c_3 \in Lib_{TetR}$$

where

$$X(c,t)=\begin{bmatrix} x_1(c_1,t) \\ X_1(c_1,t) \\ x_2(c_2,t) \\ X_2(c_2,t) \\ x_3(c_3,t) \\ X_3(c_3,t) \end{bmatrix}, f(X,c,r,t)=\begin{bmatrix} P_{Lib_{LacI}}(c_1,r_1)-\beta x_1(c_1,t) \\ \alpha x_1(c_1,t)-\gamma_1 X_1(c_1,t) \\ P_{Lib_{TetR}}(c_2,r_2)-\beta x_2(c_2,t) \\ \alpha x_2(c_2,t)-\gamma_2 X_2(c_2,t) \\ P_{Lib_{TetR}}(c_3,r_3)-\beta x_3(c_3,t) \\ \alpha x_3(c_3,t)-\gamma_3 X_3(c_3,t) \end{bmatrix}, v(t)=\begin{bmatrix} v_1(t) \\ v_2(t) \\ v_3(t) \\ v_4(t) \\ v_5(t) \\ v_6(t) \end{bmatrix}, H=\begin{bmatrix} 0 \\ 0 \\ 0 \\ 0 \\ 0 \\ 1 \end{bmatrix}^T,$$

$$M_1=\begin{bmatrix} \Delta c_{s_1} & \Delta c_{r_1} & -\Delta\beta & 0 \\ 0 & 0 & \Delta\alpha & -\Delta\gamma_1 \\ 0 & 0 & 0 & 0 \\ 0 & 0 & 0 & 0 \\ 0 & 0 & 0 & 0 \\ 0 & 0 & 0 & 0 \end{bmatrix}, M_2=\begin{bmatrix} 0 & 0 & 0 & 0 \\ 0 & 0 & 0 & 0 \\ \Delta c_{s_2} & \Delta c_{r_2} & -\Delta\beta & 0 \\ 0 & 0 & \Delta\alpha & -\Delta\gamma_2 \\ 0 & 0 & 0 & 0 \\ 0 & 0 & 0 & 0 \end{bmatrix}, M_3=\begin{bmatrix} 0 & 0 & 0 & 0 \\ 0 & 0 & 0 & 0 \\ 0 & 0 & 0 & 0 \\ 0 & 0 & 0 & 0 \\ \Delta c_{s_3} & \Delta c_{r_3} & -\Delta\beta & 0 \\ 0 & 0 & \Delta\alpha & -\Delta\gamma_3 \end{bmatrix},$$

$$g_1(X,c_1,r_1,t)=\begin{bmatrix} H_{LacI}(r_1) \\ 1-H_{LacI}(r_1) \\ x_1(c_1,t) \\ X_1(c_1,t) \end{bmatrix}, g_2(X,c_2,r_2,t)=\begin{bmatrix} H_{TetR}(r_2) \\ 1-H_{TetR}(r_2) \\ x_2(c_2,t) \\ X_2(c_2,t) \end{bmatrix}, g_3(X,c_3,r_3,t)=\begin{bmatrix} H_{TetR}(r_3) \\ 1-H_{TetR}(r_3) \\ x_3(c_3,t) \\ X_3(c_3,t) \end{bmatrix}$$

For the convenience of analysis and design, the stochastic synthetic gene network in (6.16), can be represented by the following Ito stochastic differential equation (Chen and Hsu 1995, Zhang and Chen 2006, Wu et al. 2011b)

$$dX(c,t)=\left(f(X,c,r,t)+v(t)\right)dt+\sum_{k=1}^{3} M_k g_k\left(X,c_k,r_k,t\right)d\omega_k(t)$$

$$Y(c,t)=HX(c,t) \tag{6.17}$$

$$c=\{c_1,c_2,c_3\}, c_1\in Lib_{LacI}, c_2,c_3\in Lib_{TetR}$$

The stochastic part of intrinsic parameter fluctuations is absorbed to $n_k(t)$ with $d\omega_k(t)=n_k(t)dt$, where $\omega_k(t)$ denotes a standard Wiener process or Brownian motion (Chen and Hsu 1995, Zhang and Chen 2006). Since the stochastic differential equation in (6.17) is dependent on the selection of the promoter set c from the redefined promoter libraries, the synthetic gene network with desired behaviors can be achieved by selecting adequate promoter set c.

Now, consider a more general design case of the synthetic gene network with n genes as follows (Wu et al. 2011b)

$$dX(c,t) = \left(f(X,c,r,t) + v(t)\right)dt + \sum_{k=1}^{n} M_k g_k\left(X, c_k, r_k, t\right) d\omega_k$$

$$Y(c,t) = HX(c,t) \tag{6.18}$$

$$c = \left\{c_1, c_2, \ldots, c_n\right\} \in Lib_j, \; j = 1, 2, \ldots, m$$

where the state vector $X(c,t) = \left[x_1(c_1,t), X_1(c_1,t), \ldots, x_n(c_n,t), X_n(c_n,t)\right]^T$ denotes the concentrations of mRNAs and proteins in the synthetic gene network, and $f(X,c,r,t)$ denotes the nonlinear gene regulation function consisting of promoter set c from the redefined promoter libraries Lib_j, $j = 1, 2, \ldots, m$ and other parameters (i.e., translation rates and degradation rates of mRNA and proteins). The output vector $Y(c,t) = \left[y_1(c,t), \ldots, y_l(c,t)\right]^T$ denotes the concentrations of the observed proteins of interest. H is a $l \times 2n$ matrix, where l is the number of observed proteins of interest. The elements of the perturbative matrix M_k denote the corresponding perturbation amplitudes or the standard deviations of the corresponding stochastic parameter fluctuations due to n random noises. $v(t)$ denotes the environmental disturbances. For the stochastic differential equation (6.18), we assume the equilibrium point (phenotype) of interest is at origin. If the equilibrium point of interest is not at origin, it can be shifted to the origin.

In order to engineer a stochastic synthetic gene network to robustly and optimally track a desired trajectory by selecting an adequate promoter set from the redefined promoter libraries, two design objectives are provided for the multiobjective H$_2$/H$_\infty$ reference tracking design from a systematic point of view.

Multiobjective H$_2$/H$_\infty$ reference tracking design can be achieved by selecting an adequate promoter set from the redefined promoter libraries to engineer a synthetic gene network to track the desired behavior. Initially, in order to attenuate the effects of intrinsic parameter fluctuations and environmental disturbances *in vivo*, we consider a desired noise attenuation level as the first design objective. Then, in order to minimize the tracking error between the synthetic gene network and reference model, the optimal reference tracking is considered as the second design objective. Therefore, our design purpose is to engineer a synthetic gene network with desired behavior by selecting an adequate promoter set from the redefined promoter libraries to satisfy two design objectives, i.e., the desired noise attenuation level and optimal reference tracking. Based on the real biological circuit topology and its stochastic differential equation (6.18), how to engineer a synthetic gene network with some desired behaviors becomes how to select an adequate promoter set from the redefined promoter libraries to track the reference trajectories generated by the following reference model

$$\dot{X}_r(t) = A_r X_r(t) + r(t)$$
$$Y_r(t) = H_r X_r(t) \tag{6.19}$$

where A_r and $r(t)$ are specified before hand by the designer to generate a desired behavior, $Y_r(t)$, to be tracked by $Y(c,t)$ in (6.18); and H_r is a $l \times n$ matrix, where l is the number of observed proteins. Based on these two design objectives, the stochastic differential equation (6.18) could be designed so that $Y(c,t)$ could track the desired $Y_r(t)$ generated by the reference model in (6.19) despite intrinsic parameter fluctuations and environmental disturbances. From the engineering point of view, the multiobjective H_2/H_∞ reference tracking design is specified as follows: a desired noise attenuation level ρ_d and the optimal reference tracking, i.e., the following two design objectives need to be achieved simultaneously (Chen et al. 2000, Chen and Zhang 2004, Wu et al. 2011b)

i) H_∞ noise attenuation level:

$$\frac{E \int_0^{t_f} \left(Y(c,t) - Y_r(t)\right)^T Q\left(Y(c,t) - Y_r(t)\right) dt}{E \int_0^{t_f} v^T(t) v(t) dt} \le \rho_d^2 \text{ for } c \in Lib_j, \, j = 1, \ldots, m \tag{6.20}$$

ii) H_2 optimal reference tracking:

$$\min_{\substack{c \in Lib_j \\ j=1,\ldots,m}} E \int_0^{t_f} \left(Y(c,t) - Y_r(t)\right)^T Q\left(Y(c,t) - Y_r(t)\right) dt \tag{6.21}$$

where $Y(c,t)$ and $Y_r(t)$ are the system outputs in (6.18) and (6.19), respectively; and Q is a symmetric weighting matrix. The physical meaning of the H_∞ noise attenuation level ρ_d in (i) is that the effect of all possible environmental disturbances on the tracking error $Y(c,t) - Y_r(t)$ should be less than the desired noise attenuation level ρ_d from the average energy point of view. If the desired ρ_d in (6.20) holds, the noise filtering ability of synthetic gene networks will be better than a desired noise attenuation level ρ_d, despite the intrinsic parameter fluctuations and environmental disturbances from the average energy perspective. The optimal reference tracking performance in (ii) is to select a promoter set from the redefined promoter libraries to achieve the minimum mean square tracking error. Hence if these two design objectives are satisfied, then the synthetic gene network can simultaneously achieve both robustness and optimal reference tracking objectives. In summary, the multiobjective H_2/H_∞ reference tracking design is to select an adequate promoter set $c = \{c_1, c_2, \ldots c_n\}$ from the redefined promoter libraries Lib_j, $j = 1, 2, \ldots, m$ so that the desired noise attenuation level ρ_d

in (6.20) and the optimal reference tracking in (6.21) for the synthetic gene network are achieved simultaneously.

Based on the analysis above, the design steps of multiobjective H_2/H_∞ reference tracking for the synthetic gene network are outlined as follows (Wu et al. 2011b). (i) Based on the prescribed gene circuit topology as a guide, the dynamic models of the synthetic gene network and reference model should be constructed at first. (ii) The desired standard deviations of parameter fluctuations M_k in (6.18) to be tolerated by the stochastic gene network *in vivo* are thus specified. (iii) An adequate promoter set $c = \{c_1, c_2, \ldots, c_n\}$ is selected from promoter libraries Lib_j, $j = 1, 2, \ldots, m$ for a synthetic gene network to simultaneously satisfy the desired noise attenuation level ρ_d in (6.20) and optimal reference tracking in (6.21). Therefore, following the above design steps, the synthetic gene network with some desired behaviors can be achieved without the large number of trial-and-error experiments in conventional methods.

Our design purpose is to select an adequate promoter set $c = \{c_1, c_2, \ldots c_n\}$ for the synthetic gene network from promoter libraries Lib_j, $j = 1, 2, \ldots, m$ to satisfy the multiobjective H_2/H_∞ design objectives in (6.20) and (6.21). To illustrate the design procedure, we combine the stochastic gene network in (6.18) with the reference model in (6.19) as an augmented system

$$
\begin{bmatrix} dX(c,t) \\ dX_r(t) \end{bmatrix} = \left(\begin{bmatrix} f(X,c,r,t) \\ A_r X_r(t) \end{bmatrix} + \begin{bmatrix} v(t) \\ r(t) \end{bmatrix} \right) dt + \sum_{k=1}^{n} \begin{bmatrix} M_k g_k(X, c_k, r_k, t) \\ 0 \end{bmatrix} d\omega_k
$$

$$
\begin{bmatrix} Y(c,t) \\ Y_r(t) \end{bmatrix} = \begin{bmatrix} H & 0 \\ 0 & H_r \end{bmatrix} \begin{bmatrix} X(c,t) \\ X_r(t) \end{bmatrix}
$$

(6.22)

or equivalently,

$$
d\overline{X}(c,t) = \left(\overline{f}(\overline{X}, c, r, t) + \overline{v}(t) \right) dt + \sum_{k=1}^{n} \overline{M}_k \overline{g}_k(\overline{X}, c_k, r_k, t) d\omega_k
$$

$$
\overline{Y}(c,t) = \overline{H}\overline{X}(c,t)
$$

(6.23)

where $\overline{X}(c,t) = \begin{bmatrix} X(c,t) \\ X_r(t) \end{bmatrix}$, $\overline{Y}(c,t) = \begin{bmatrix} Y(c,t) \\ Y_r(t) \end{bmatrix}$,

$\overline{v} = \begin{bmatrix} v(t) \\ r(t) \end{bmatrix}$, $\overline{f}(\overline{X}, c, r, t) = \begin{bmatrix} f(X, c, r, t) \\ A_r X_r(t) \end{bmatrix}$, $\overline{M}_k = \begin{bmatrix} M_k & 0 \end{bmatrix}^T$,

$\overline{g}_k(\overline{X}, c_k, r_k, t) = g_k(X, c_k, r_k, t)$ and $\overline{H} = diag(H, H_r)$. Then the two design objectives in (6.20) and (6.21) are equivalent to the following:

$$\frac{E\int_0^{t_f} \overline{Y}^T(c,t)\overline{Q}\,\overline{Y}(c,t)\,dt}{E\int_0^{t_f} \overline{v}^T(t)\overline{v}(t)\,dt} \leq \rho_d^2 \text{ or } E\int_0^{t_f} \overline{Y}^T(c,t)\overline{Q}\,\overline{Y}(c,t)\,dt \leq \rho_d^2 E\int_0^{t_f} \overline{v}^T(t)\overline{v}(t)\,dt \quad (6.24)$$

and

$$\min_{\substack{c\in Lib_j \\ j=1,\ldots,m}} E\int_0^{t_f} \overline{Y}^T(c,t)\overline{Q}\overline{Y}(c,t)\,dt \quad\quad\quad (6.25)$$

respectively, where $\overline{Q} = \begin{bmatrix} Q & -Q \\ -Q & Q \end{bmatrix}$.

Remark 6.1: If the initial value $\overline{X}(c,0) \neq 0$, then the inequality in (6.24) should be modified as

$$E\int_0^{t_f} \overline{Y}(c,t)\overline{Q}\overline{Y}(c,t)\,dt \leq V\left(\overline{X}(c,0)\right) + \rho_d^2 E\int_0^{t_f} \overline{v}^T(t)\overline{v}(t)\,dt \text{ for some}$$

Lyapunov functions $V(\overline{X}(c,t)) > 0$, i.e., the effect of initial values of system states should be taken into consideration.

Based on the augmented system in (6.22) or (6.23), our design purpose is to select some adequate promoter sets $c = \{c_1, c_2, \ldots, c_n\}$ from the redefined promoter libraries Lib_j, $j = 1, 2, \ldots, m$ such that the desired noise attenuation level ρ_d in (6.24) can be achieved. Then from these adequate promoter sets, we select one promoter set to achieve the optimal reference tracking in (6.25). In order to achieve the above two design objectives by a systematic method, design step (iii) can be divided into two steps. In the first step, we select all possible promoter sets from the redefined promoter libraries to satisfy the desired noise attenuation level ρ_d in (6.24), though there may exist several promoter sets satisfying the desired noise attenuation level ρ_d in (6.24). Then our second step is to select a promoter set from these promoter sets, which satisfies the desired noise attenuation level, to achieve the optimal reference tracking in (6.25). In this case, we choose a Lyapunov (energy) function $V(\overline{X}) > 0$ with $V(0) = 0$ for the augmented stochastic gene network in (6.23). Based on the Lyapunov function, we obtain the following result.

Proposition 6.1

For the stochastic gene network in (6.18), if some promoter sets $c = \{c_1, c_2, \ldots, c_n\}$ are selected from the redefined promoter libraries Lib_j, $j = 1, 2, \ldots m$ so that the following Hamilton-Jacobi inequality (HJI) has a positive solution $V(\overline{X}) > 0$ for each promoter set

$$\left(\frac{\partial V\left(\bar{X}(c,t)\right)}{\partial \bar{X}}\right)^{T}\bar{f}(\bar{X},c,r,t)+\frac{1}{2}\sum_{k=1}^{n}\bar{g}_{k}^{T}(\bar{X},c_{k},r_{k},t)\bar{M}_{k}^{T}\frac{\partial^{2}V\left(\bar{X}(c,t)\right)}{\partial \bar{X}^{2}}\bar{M}_{k}\bar{g}_{k}(\bar{X},c_{k},r_{k},t)$$

$$+\frac{1}{4\rho_{d}^{2}}\left(\frac{\partial V\left(\bar{X}(c,t)\right)}{\partial \bar{X}}\right)^{T}\left(\frac{\partial V\left(\bar{X}(c,t)\right)}{\partial \bar{X}}\right)+\bar{X}^{T}(c,t)\bar{H}^{T}\bar{Q}\bar{H}\bar{X}(c,t)\le 0 \tag{6.26}$$

then the synthetic gene network with these promoter sets have a desired noise attenuation level ρ_{d}.

Proof: See Appendix 6.1.

There may exist several promoter sets selected from the redefined promoter libraries $Lib_{j}, j = 1, 2, \ldots m$, which could solve HJI in (6.26) for the synthetic gene network with a desired noise attenuation level ρ_{d}. Our second step is to select a promoter set that has an H_{∞} noise attenuation level from these candidate promoter sets to achieve the H_{2} optimal reference tracking in (6.25). Based on the suboptimal tracking design, we can obtain the following result.

Proposition 6.2

The suboptimal tracking control design problem in (6.25) for synthetic gene networks in (6.18) becomes how to select one promoter set to solve the following HJI-constrained optimization problem

$$\min_{\substack{c\in Lib_{j}\\ j=1,\ldots,m}} EV\left(\bar{X}(c,0)\right)$$

$$\text{subject to } \left(\frac{\partial V\left(\bar{X}(c,t)\right)}{\partial \bar{X}}\right)^{T}\bar{f}(\bar{X},c,r,t)+\frac{1}{2}\sum_{k=1}^{n}\bar{g}_{k}^{T}(\bar{X},c_{k},r_{k},t)\bar{M}_{k}^{T}\frac{\partial^{2}V\left(\bar{X}(c,t)\right)}{\partial \bar{X}^{2}}\bar{M}_{k}\bar{g}_{k}(\bar{X},c_{k},r_{k},t) \tag{6.27}$$

$$+\frac{1}{4}\left(\frac{\partial V\left(\bar{X}(c,t)\right)}{\partial \bar{X}}\right)^{T}\left(\frac{\partial V\left(\bar{X}(c,t)\right)}{\partial \bar{X}}\right)+\bar{X}^{T}(c,t)\bar{H}^{T}\bar{Q}\bar{H}\bar{X}(c,t)\le 0$$

where $V\left(\bar{X}(c,t)\right) > 0$ is the Lyapunov function of nonlinear stochastic gene networks in (6.18).

Proof: See Appendix 6.2.

In order to simultaneously satisfy the two design objectives in (6.24) and (6.25), a promoter set should be selected so that (6.26) and (6.27) simultaneously hold, i.e., we need to select a promoter set to solve the following HJI-constrained optimization problem

$$\min_{\substack{c\in Lib_{j}\\ j=1,\ldots,m}} EV\left(\bar{X}(c,0)\right) \tag{6.28}$$

subject to (6.26) and (6.27)

From the analysis above, engineering a synthetic gene network to achieve the multiobjective H_2/H_∞ reference tracking design becomes how to select an adequate promoter set $c = \{c_1, c_2, \ldots, c_n\}$ from the redefined promoter libraries Lib_j, $j = 1, 2, \ldots, m$ to solve the HJI-constrained optimization in (6.28), i.e., the two-step design procedures in Proposition 6.1 for H_∞ noise filtering and Proposition 6.2 for H_2 optimal tracking are merged in an one-step design procedure in (6.28) for solving the H_2 optimal tracking and H_∞ noise filtering problems simultaneously by only selecting one promoter set from promoter libraries. But it is still generally very difficult to solve the above HJI-constrained optimization due to the highly nonlinear properties of synthetic gene networks since there is still no efficient analytic or numerical method to solve the HJI-constrained optimization problem in (6.28).

Recently, however, the T-S fuzzy model has been widely applied to approximate the nonlinear system via interpolating several linearized systems at different operation points. Suppose the nonlinear gene network in (6.18) can be represented by the T-S fuzzy model (Takagi and Sugeno 1985), which is a piece wise interpolation of several local linearized models through the membership functions. The fuzzy model is described by several if-then rules and can be employed to approximate nonlinear gene network. The ith rule of the fuzzy model for the nonlinear stochastic system in (6.18) can be expressed as the following form (Takagi and Sugeno 1985)

Rule i:

If $X_1(c_1, t)$ is F_{1i} and ... and $X_n(c_n, t)$ is F_{ni}, then

$$dX(c,t) = \left(A_i(c)X(c,t) + v(t)\right) dt + \sum_{k=1}^{n} M_k B_{ik}(c)X(c_k, t) d\omega_k, \quad i = 1, \ldots, L \quad (6.29)$$

where F_{ji} is the fuzzy set, and L is the number of if-then rules. Then the fuzzy system in (6.29) can be inferred as follows (Wu et al. 2011b):

$$
\begin{aligned}
dX(c,t) &= \frac{\sum_{i=1}^{L} \mu_i(X)\left[\left(A_i(c)X(c,t) + v(t)\right) dt + \sum_{k=1}^{n} M_k B_{ik}(c)X(c_k, t) d\omega_k\right]}{\sum_{i=1}^{L} \mu_i(X)} \\
&= \sum_{i=1}^{L} h_i(X)\left[\left(A_i(c)X(c,t) + v(t)\right) dt + \sum_{k=1}^{n} M_k B_{ik}(c)X(c_k, t) d\omega_k\right]
\end{aligned}
\quad (6.30)
$$

where $\mu_i(X)=\prod_{j=1}^{n}F_{ji}\left(X_j\left(c_j,t\right)\right)$, $h_i(X)=\mu_i(X)/\sum_{i=1}^{L}\mu_i(X)$, and $F_{ji}(X_j(c_j,t))$ is the grade of the membership function of X_j in F_{ij}. We assume $\mu_i(X)>0$ and $\sum_{i=1}^{L}\mu_i(X)>0, \forall t$. Therefore, we obtain the fuzzy bases as $h_i(X)\geq 0$ and $\sum_{i=1}^{L}h_i(X)=1, \forall t$.

The T-S fuzzy model in (6.30) is to interpolate L linear local stochastic dynamic systems to approximate the stochastic nonlinear gene network in (6.18) via the fuzzy basis function $h_i(X(c,t))$. The matrices $A_i(c)$ and $B_{ik}(c)$, $i=1, 2, \ldots, L$, are specified so that $\sum_{i=1}^{L}h_i(X)A_i(c)X(c,t)$ and $\sum_{i=1}^{L}h_i(X)B_{ik}(c)X(c,t)$ in (6.30) can approximate $f(X,c,r,t)$ and $g_k(X,c,r,t)$ in (6.23) by the fuzzy identification method (Takagi and Sugeno 1985), respectively. Since this section focuses on the topic of multiobjective H_2/H_∞ synthetic gene network design, for simplicity, the fuzzy approximation error is neglected in (6.30) and is merged in the disturbances $v(t)$. Then the augmented system can be approximated by

$$d\bar{X}(c,t)=\sum_{i=1}^{L}h_i(X)\left[\left(\bar{A}_i(c).\bar{X}(c,t)+\bar{v}(t)\right)dt+\sum_{k=1}^{n}\bar{M}_k\bar{B}_{ik}(c)\bar{X}(c,t)\,d\omega(t)\right]$$

$$\bar{Y}(c,t)=\bar{H}\bar{X}(c,t) \tag{6.31}$$

where $\bar{A}_i(c)=diag(A_i(c),A_r)$, $\bar{M}_k=[M_k\quad 0]^T$, $\bar{B}_{ik}(c)=[B_{ik}(c)\quad 0]$.

After approximating the stochastic gene network in (6.18) by using the T-S fuzzy system in (6.30), the stochastic nonlinear multiobjective H_2/H_∞ reference tracking problem can be replaced by solving the fuzzy stochastic multiobjective H_2/H_∞ reference tracking problem in (6.24) and (6.25) for the fuzzy system in (6.31). We could choose a Lyapunov (energy) function $V(\bar{X}(c,t))=\bar{X}(c,t)^T P\bar{X}(c,t)>0$ for the fuzzy system in (6.31). Then the multiobjective H_2/H_∞ reference tracking design problem for synthetic gene networks becomes how to select a promoter set from promoter libraries Lib_j, $j=1,2,\ldots,m$ to solve the following constrained optimization problem.

Proposition 6.3

Based on T-S fuzzy approximation model in (6.31), if we could select a promoter set from the redefined promoter libraries Lib_j, $j=1,2,\ldots,m$ to solve the following linear matrix inequality (LMI)-constrained optimization

$$\min_{\substack{c \in Lib_j \\ j=1,2,\ldots,m}} PR_0$$

subject to $P > 0$

$$\begin{bmatrix} \bar{A}_i^T(c)P + P\bar{A}_i(c) + \bar{H}^T\bar{Q}\bar{H} + \sum_{k=1}^n \bar{B}_{ik}^T(c)\bar{M}_k^T P\bar{M}_k\bar{B}_{ik}(c) & P \\ P & -\rho_d^2 \end{bmatrix} < 0 \quad (6.32)$$

$$\begin{bmatrix} \bar{A}_i^T(c)P + P\bar{A}_i(c) + \bar{H}^T\bar{Q}\bar{H} + \sum_{k=1}^n \bar{B}_{ik}^T(c)\bar{M}_k^T P\bar{M}_k\bar{B}_{ik}(c) & P \\ P & -I \end{bmatrix} < 0$$

where $R_0 = E\{\bar{X}(0)\bar{X}^T(0)\}$ denotes the covariance matrix of initial condition $\bar{X}(0)$, then the H_2/H_∞ reference tracking in (6.24) and (6.25) can be achieved.

Proof: See Appendix 6.3.

The LMIs-constrained optimization problem for multiobjective H_2/H_∞ reference tracking design of the synthetic gene network in (6.32) could be efficiently solved by selecting an adequate promoter set from the redefined promoter libraries with the help of the LMI toolbox in Matlab.

Based on the analysis above, the design procedure is summarized as follows (Wu et al. 2011b).

1) Create a synthetic gene circuit topology (e.g., Figure 6.5) as a guide with promoters to be selected from the refined promoter libraries in Table 5.1, and then the stochastic differential equation in (6.18) is constructed for the synthetic gene network.
2) Provide a desired reference model in (6.19) to be tracked.
3) Provide the standard deviations of parameter fluctuations M_k in (6.18) to be tolerated by the synthetic gene network *in vivo*.
4) Construct a T-S fuzzy model in (6.30) to approximate the synthetic gene network in (6.18); and then the augmented system in (6.31), combining the T-S fuzzy model in (6.30) with the reference model in (6.19), is constructed.
5) Provide the desired H_∞ noise attenuation level ρ_d.
6) Solve the LMI-constrained optimization problem in (6.32) by selecting a promoter set via searching algorithm from the redefined promoter libraries.

A simple flowchart is shown in Figure 6.6.

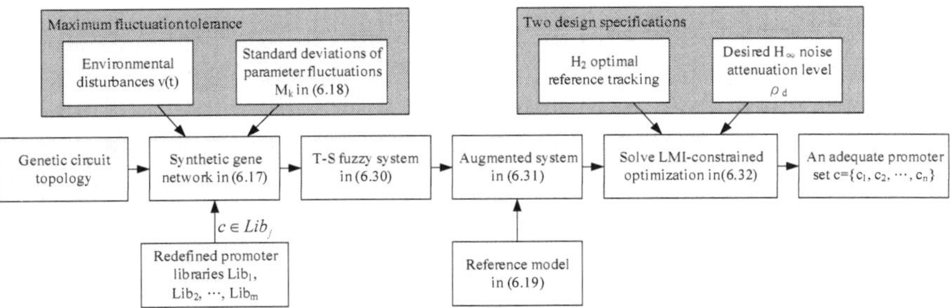

Figure 6.6. Flowchart of design procedure for multiobjective H_2/H_∞ synthetic gene network.

6.2.2 In silico design examples

In this section, we provide two *in silico* examples to illustrate the design procedure of synthetic gene networks, and then an adequate promoter set is selected from the redefined promoter libraries Lib_{TetR} and Lib_{LacI} in Table 5.1 to achieve the multiobjective H_2/H_∞ reference tracking design. The redefined promoter libraries Lib_{TetR} and Lib_{LacI} are listed in Table 5.1, and all design parameters for simulation are listed in Table 5.2.

6.2.2.1 Design of robust and optimal synthetic toggle switch by multiobjective H_2/H_∞ tracking performance

Consider the synthetic toggle switch shown in Figure 6.5. The gene network will employ adequate promoters from TetR- and LacI-regulated promoter libraries (i.e., Lib_{TetR} and Lib_{LacI}), which have both 21 promoters $T_0 - T_{20}$ and $L_0 - L_{20}$ in Table 5.1. Under intrinsic parameter fluctuations and environmental disturbances, the dynamic system in Figure 6.5 has been constructed in (6.16), where $c_1 = \{c_{r_1}, c_{s_1}\} \in L_k$, $c_2 = \{c_{r_2}, c_{s_2}\} \in T_k$, $c_3 = \{c_{r_3}, c_{s_3}\} \in T_k$, $k = 0,\dots,20$. The elements of parameter fluctuation matrices M_k, $k = 1, 2, 3$ in (6.16) are given as follows

$$\Delta c_{r_k} = 0.1 \times c_{r_k},\ \Delta c_{s_k} = 0.1 \times c_{s_k},\ k = 1, 2, 3$$

$$\Delta \beta = 0.1 \times \beta,\ \Delta \alpha = 0.1 \times \alpha,\ \Delta \gamma = 0.1 \times \gamma \tag{6.33}$$

i.e., the standard deviations of parameter fluctuations are allowed to be 10% of their nominal values. Finally, the reference model to be tracked by the stochastic synthetic gene network in Figure 6.5 is given as

$$\dot{X}_r(t) = -0.0019 X_r(t) + r$$

$$Y_r(t) = X_r(t) \tag{6.34}$$

where $r = 60000$ if ATc induces or $r = 2000$ if IPTG induces (see Figure 6.7). For simplicity, the stochastic synthetic gene network in Figure 6.5 is approximated by the T-S fuzzy model. We take 5 triangle-type membership functions for Rule 1 to Rule 5 (i.e., 25 local linearized dynamic systems) and the operating points for all states are all distributed from 0 to 1000. The parameters for simulation are listed in Table 5.2. Then the augmented fuzzy system in (6.31) is obtained as follows:

$$d\bar{X}(c,t) = \sum_{i=1}^{25} h_i(\bar{X}) \left[\left(\bar{A}_i(c)\bar{X}(c,t) + \bar{v}(t) \right) dt + \sum_{k=1}^{3} \bar{M}_k \bar{B}_{ik}(c)\bar{X}(c,t)\, d\omega_k \right] \quad (6.35)$$

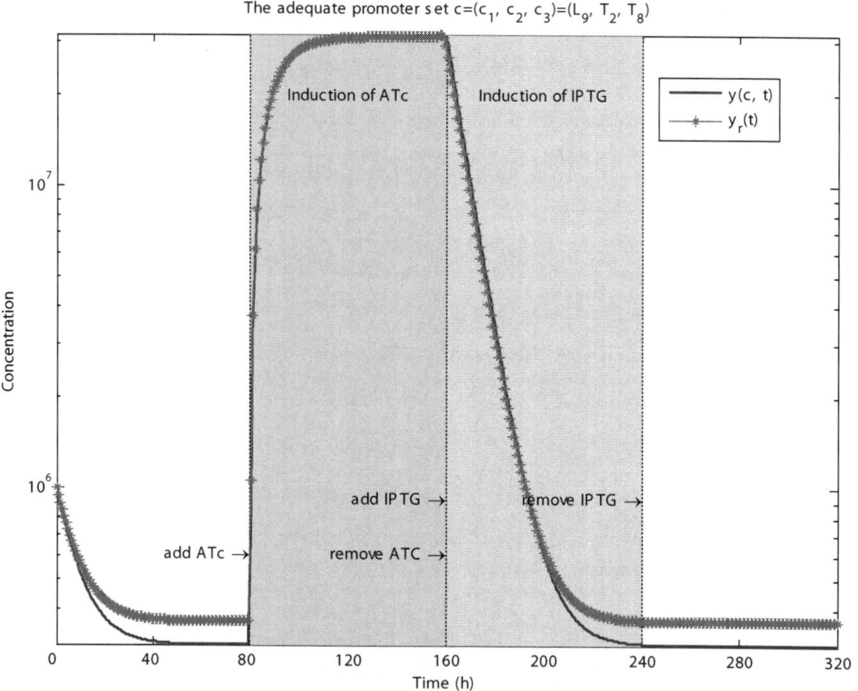

Figure 6.7. Simulation results for synthetic toggle switch design. By solving the LMIs-constrained optimization problem in (6.32) for the synthetic gene network in Figure 6.5, an adequate promoter set $c = \{c_1, c_2, c_3\} = \{L_9, T_2, T_8\}$ is selected from the corresponding promoter libraries. The synthetic gene network is added with inducer ATc to induce the gene network from 80 hours to 160 hours, and then is added with inducer IPTG from 160 hours to 240 hours. Obviously, the output $Y(c,t)$ can robustly track the desired reference output $Y_r(t)$.

Color image of this figure appears in the color plate section at the end of the book.

where $\overline{A}_i(c)$ and $\overline{B}_{ik}(c)$ are obtained by the fuzzy approximating method and $\overline{M}_k$ is shown in (6.33). Assume the weighing matrix $Q=1$, the covariance of initial value $R_0 = \overline{X}(0)\overline{X}^T(0)$ with $\overline{X}(0)=10^3 \times [1 \quad 1 \quad 1 \quad 1 \quad 1 \quad 10^3 \quad 10^3]^T$ and the desired H_∞ noise attenuation level ρ_d is 0.1, then the LMIs-constrained optimization problem in (6.32) is solved by selecting a promoter set from the redefined promoter libraries. With the help of the LMI toolbox in Matlab, the adequate promoter set $c = \{c_1, c_2, c_3\} = \{L_9, T_2, T_8\}$ from the corresponding promoter libraries can be selected so that the synthetic gene network in Figure 6.5 can achieve multiobjective H_2/H_∞ reference tracking in (6.32). For the convenience of simulation, we consider the environmental disturbances $v(t)=10 \times [n_1, n_2, \ldots, n_6]^T$, where $n_i, i=1,\ldots,6$ are independent Gaussian white noises with zero mean and unit variance. In order to confirm the multiobjective reference tracking, the inducer ATc is added to induce the gene network from 80 hours and is removed at 160 hours, and then the inducer IPTG is added to the synthetic gene network from 160 hours and is removed at 240 hours. The simulation results for reference tracking of the synthetic gene networks are shown in Figure 6.7. Clearly, the output $Y(c,t)$ of the synthetic gene network can robustly and optimally track the desired reference $Y_r(t)$ despite the intrinsic parameter fluctuations and environmental disturbances.

6.2.2.2 Design of robust and optimal synthetic genetic oscillator by multiobjective H_2/H_∞ tracking performance

Consider the synthetic genetic oscillator shown in Figure 6.3. The nonlinear stochastic model with the four promoters c_1, c_2, c_3 and c_4 to be specified under intrinsic parameter fluctuations and environmental disturbances is given as

$$dX(c,t) = \left(f(X,c,r,t) + v(t)\right)dt + \sum_{k=1}^{4} M_k g_k(X,c_k,r_k,t)\, d\omega_k$$

$$y(C,t) = X_4(c_4,t)$$

$$(6.36)$$

where

$$X(c,t)=\begin{bmatrix} x_1(c_1,t) \\ X_1(c_1,t) \\ x_2(c_2,t) \\ X_2(c_2,t) \\ x_3(c_3,t) \\ X_3(c_3,t) \\ x_4(c_4,t) \\ X_4(c_4,t) \end{bmatrix}, f(X,c,r,t)=\begin{bmatrix} p_{Lib_{TetR}}(c_1,r_1)-\beta x_1(c_1,r_1) \\ \alpha x_1(c_1,t)-\gamma_1 X_1(c_1,t) \\ p_{Lib_{LacI}}(c_2,r_2)-\beta x_2(c_2,r_2) \\ \alpha x_2(c_2,t)-\gamma_2 X_2(c_2,t) \\ p_{Lib_{CI}}(c_3,r_3)-\beta x_3(c_3,r_3) \\ \alpha x_3(c_3,t)-\gamma_3 X_3(c_3,t) \\ p_{Lib_{TetR}}(c_4,r_4)-\beta x_4(c_4,r_4) \\ \alpha x_4(c_4,t)-\gamma_4 X_4(c_4,t) \end{bmatrix}, v(t)=\begin{bmatrix} v_1(t) \\ v_2(t) \\ v_3(t) \\ v_4(t) \\ v_5(t) \\ v_6(t) \\ v_7(t) \\ v_8(t) \end{bmatrix},$$

$$M_1=\begin{bmatrix} \Delta c_{s_1} & \Delta c_{r_1} & -\Delta\beta & 0 \\ 0 & 0 & \Delta\alpha & -\Delta\gamma_1 \\ 0 & 0 & 0 & 0 \\ 0 & 0 & 0 & 0 \\ 0 & 0 & 0 & 0 \\ 0 & 0 & 0 & 0 \\ 0 & 0 & 0 & 0 \\ 0 & 0 & 0 & 0 \end{bmatrix}, M_2=\begin{bmatrix} 0 & 0 & 0 & 0 \\ 0 & 0 & 0 & 0 \\ \Delta c_{s_2} & \Delta c_{r_2} & -\Delta\beta & 0 \\ 0 & 0 & \Delta\alpha & -\Delta\gamma_2 \\ 0 & 0 & 0 & 0 \\ 0 & 0 & 0 & 0 \\ 0 & 0 & 0 & 0 \\ 0 & 0 & 0 & 0 \end{bmatrix},$$

$$M_3=\begin{bmatrix} 0 & 0 & 0 & 0 \\ 0 & 0 & 0 & 0 \\ 0 & 0 & 0 & 0 \\ 0 & 0 & 0 & 0 \\ \Delta c_{s_3} & \Delta c_{r_3} & -\Delta\beta & 0 \\ 0 & 0 & \Delta\alpha & -\Delta\gamma_3 \\ 0 & 0 & 0 & 0 \\ 0 & 0 & 0 & 0 \end{bmatrix}, M_4=\begin{bmatrix} 0 & 0 & 0 & 0 \\ 0 & 0 & 0 & 0 \\ 0 & 0 & 0 & 0 \\ 0 & 0 & 0 & 0 \\ 0 & 0 & 0 & 0 \\ 0 & 0 & 0 & 0 \\ \Delta c_{s_4} & \Delta c_{r_4} & -\Delta\beta & 0 \\ 0 & 0 & \Delta\alpha & -\Delta\gamma_4 \end{bmatrix},$$

$$g_1(X,c_1,r_1,t) = \begin{bmatrix} H_{TetR}(r_1) \\ 1 - H_{TetR}(r_1) \\ x_1(c_1,t) \\ X_1(c_1,t) \end{bmatrix}, \; g_2(X,c_2,r_2,t) = \begin{bmatrix} H_{LacI}(r_2) \\ 1 - H_{LacI}(r_2) \\ x_2(c_2,t) \\ X_2(c_2,t) \end{bmatrix},$$

$$g_3(X,c_3,r_3,t) = \begin{bmatrix} H_{CI}(r_3) \\ 1 - H_{CI}(r_3) \\ x_3(c_3,t) \\ X_3(c_3,t) \end{bmatrix}, \; g_4(X,c_4,r_4,t) = \begin{bmatrix} H_{TetR}(r_4) \\ 1 - H_{TetR}(r_4) \\ x_4(c_4,t) \\ X_4(c_4,t) \end{bmatrix}$$

with $r_1 = X_3(c_3,t)$, $r_2 = X_1(c_1,t)$, $r_3 = X_2(c_2,t)$ and $r_4 = X_3(c_3,t)$.

Suppose the perturbation elements $\Delta\beta$, $\Delta\alpha$, $\Delta\gamma$, Δc_{s_k} and Δc_{r_k}, $k = 1, \ldots, 4$ of parameter fluctuation matrices M_k in (6.36) to be tolerated are listed as follows

$$\Delta c_{r_k} = 0.1 \times c_{r_k}, \; \Delta c_{s_k} = 0.1 \times c_{s_k}, \; k = 1, \ldots, 4$$
$$\Delta\beta = 0.1 \times \beta, \; \Delta\alpha = 0.1 \times \alpha, \; \Delta\gamma = 0.1 \times \gamma, \tag{6.37}$$

Since there is currently no CI-regulated promoter library, the promoter regulation function of c_3 takes the following form (Basu et al. 2005)

$$p_{CI}(c_3,r_3) = 150 / (1 + (r_3 / 20)^2) \tag{6.38}$$

The promoters c_1, c_2 and c_4 are to be selected from promoters T_k, L_k and T_k, $k = 0, 1, \ldots, 20$ in Table 5.1 in promoter libraries Lib_{TetR}, Lib_{LacI} and Lib_{TetR}, respectively.

Our design purpose is to engineer a synthetic gene network whose behavior can achieve the multiobjective H_2/H_∞ reference tracking for the following reference oscillation system with the desired amplitude 400000 (M) and frequency 8×10^{-3} (min^{-1}) as

$$\begin{bmatrix} \dot{X}_{r1}(t) \\ \dot{X}_{r2}(t) \\ \dot{X}_{r3}(t) \end{bmatrix} = \begin{bmatrix} 0 & 0.029 & -0.029 \\ -0.029 & 0 & 0.029 \\ 0.029 & -0.029 & 0 \end{bmatrix} \begin{bmatrix} X_{r1}(t) \\ X_{r2}(t) \\ X_{r3}(t) \end{bmatrix} \tag{6.39}$$

Take triangle-type membership functions for Rule 1 to Rule 5 and the operating points for all states are distributed from 0 to 1000. Then the augmented system combining the T-S fuzzy model with the reference model in (6.39) is obtained as

$$d\bar{X}(c,t)= \sum\nolimits_{i=1}^{125} h_i\left(\bar{X}\right)\left[\left(\bar{A}_i(c)\bar{X}(c,t)+\bar{v}(t)\right)dt + \sum_{k=1}^{3}\bar{M}_k\bar{B}_{ik}(c)\bar{X}(c,t)\,d\omega_k\right] \quad (6.40)$$

where $\bar{A}_i(c)$ and $\bar{B}_{ik}(c)$ are obtained by the fuzzy approximating method and $\bar{M}_k$ is shown in (6.36). Assume the weighting matrix $Q=1$, the covariance of initial value $R_0 = \bar{X}(0)\bar{X}^T(0)$ with $\bar{X}(0)=10^4\times[1\quad 1\quad 1\quad 1\quad 1\quad 1\quad 1\quad 1\quad 260\quad 200\quad 200]^T$ and the desired H_∞ noise attenuation level ρ_d is 0.1. Then the LMIs-constrained optimization problem in (6.32) is solved by selecting a promoter set from the redefined promoter libraries in Table 5.1. The adequate promoter set $c = \{c_1,c_2,c_4\} = \{T_{11},L_{15},T_8\}$ can be obtained by solving the LMIs-constrained optimization problem in (6.32) with the help of the LMI toolbox in Matlab, so that the synthetic gene network in Figure

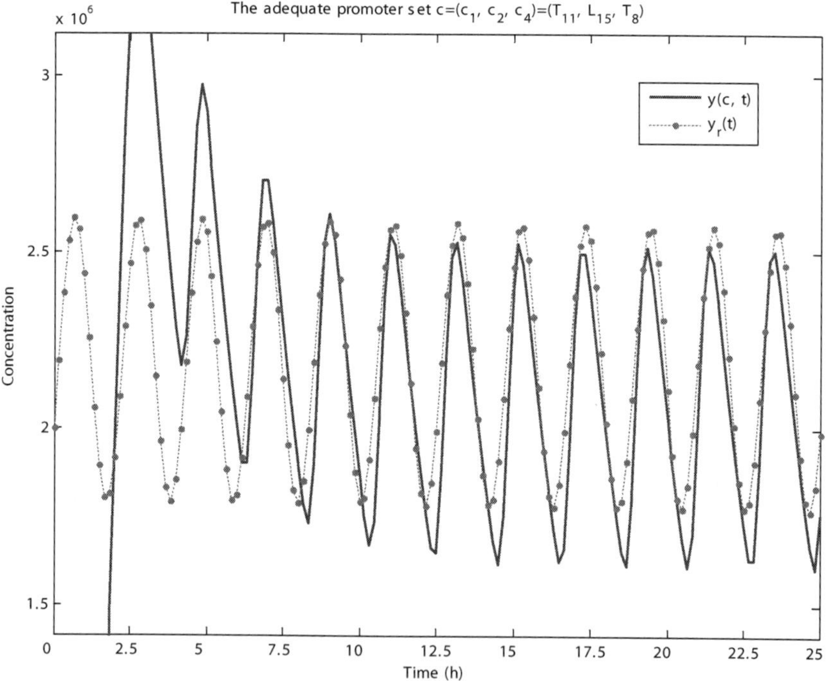

Figure 6.8. Simulation results for synthetic genetic oscillator design. The adequate promoter set $c = \{c_1,c_2,c_4\} = \{T_{11},L_{15},T_8\}$ is obtained via solving the multiobjective H_2/H_∞ reference tracking design problem in (6.32) for the synthetic gene network in Figure 6.3. The desired output $Y(c,t)$ of the gene network employs the adequate promoter set c to robustly and optimally track the desired behavior $Y_r(t)$ generated by the reference model. Our proposed design method could provide a genetic oscillator with the prescribed amplitude and period via selecting an adequate promoter set from the existing promoter libraries.

Color image of this figure appears in the color plate section at the end of the book.

6.3 can achieve multiobjective H_2/H_∞ reference tracking design. For convenience of simulation, we consider the environmental disturbances $v(t) = 10 \times [n_1, n_2, \ldots, n_8]^T$, in which $n_i(t), i = 1, \ldots, 8$ are independent Gaussian white noises with zero mean and unit variance. The simulation results in Figure 6.8 show that the gene network can robustly track the desired behavior generated by the reference model in (6.39) despite intrinsic parameter fluctuations and environmental disturbances.

6.2.3 Summary

Nowadays, promoter library studies have significant progress in quantitative measurements. Based on the knowledge about the effect of the promoter architecture on transcriptional activity and mutation-selection techniques, the promoter libraries can be widely built. Hence the proposed method can be employed to redefine these promoter libraries to extend a more complex synthetic gene network. For example, the next-generation gene networks, such as tunable filters, analog-to-digital and digital-to-analog converters, adaptive learning networks and protein-based computational circuits, have been proposed to enable the construction of more complex biological systems based on diverse biological parts (Lu et al. 2009). In the future, a much larger and more complex synthetic gene network can be represented by a complex nonlinear stochastic system in (6.18) and the desired behavior of a synthetic gene network can be described by the reference model in (6.19). Then, based on the fuzzy interpolation method in (6.31), the multiobjective H_2/H_∞ reference tracking design can be efficiently solved by Proposition 6.3. By the conventional selecting algorithm, an adequate promoter set can be efficiently searched from promoter libraries to achieve the minimum value in (6.32). When the number and content of promoter libraries become larger in the future, the library-based search method can be efficiently employed in the systematic design of synthetic biology. Therefore, one of the future challenges in synthetic biology is to engineer some application-orientated systems. After providing a novel and functional network, the proposed systematic method can be employed to implement the synthetic gene network with the desired function by selecting an adequate promoter set. By the proposed library-searching method, the engineering efforts are needed to focus on creating a systems-level network to promote the second wave of synthetic biology (Purnick and Weiss 2009). If the BioBrick parts can be characterized by their promoter activities by the proposed method of redefined promoter library in Chapter 5, then the BioBricks can be utilized as possible resources of promoter libraries for synthetic gene networks. In this situation, the proposed systematic method based on well-characterized BioBrick parts will surely shape the future design in synthetic biology.

In this section, we propose a multiobjective H_2/H_∞ tracking design for a stochastic synthetic gene network to achieve H_∞ robust noise filtering and H_2 optimal reference tracking simultaneously by selecting an adequate promoter set from the existing promoter libraries. The multiobjective H_2/H_∞ tracking design problem is transformed into an HJI-constrained optimization problem. Then the T-S fuzzy method is employed to interpolate several local linear stochastic systems for approximating the nonlinear stochastic gene network *in vivo*. Thus the HJI-constrained optimization problem is transformed to that of selecting a promoter set to solve a LMIs-constrained optimization problem, which can be efficiently solved using the LMI toolbox in Matlab. For the two design examples *in silico*, robust and optimal reference tracking of synthetic gene networks can be guaranteed simultaneously despite the intrinsic parameter fluctuations and environmental disturbances *in vivo*. Hence, the proposed systematic method will help synthetic biologists simplify the design procedure of a synthetic gene network with a desired behavior. A large number of trial-and-error experimental processes in selecting an adequate promoter set from promoter libraries can be avoided. Therefore, the proposed systematic method will accelerate the progress of synthetic biology, especially as the available biological promoter libraries increase rapidly.

Appendix 6.1: Proof of Proposition 6.1

Let us choose a Lyapunov function $V(\bar{X}(c,t)) > 0$ for $\bar{X}(c,t) \neq 0$ with $V(0) = 0$ for the synthetic gene network in (6.23). Then we obtain

$$E\int_0^{t_f} \bar{Y}^T(c,t)\overline{Q}\bar{Y}(c,t)\ dt = EV(\bar{X}(c,0)) - EV(\bar{X}(c,t_f))$$

$$+E\int_0^{t_f}\left[\bar{Y}^T(c,t)\overline{Q}\bar{Y}(c,t) + \frac{dV(\bar{X}(c,t))}{dt}\right] dt \tag{6.41}$$

By the Ito formula (Chen and Hsu 1995) and $E(d\omega_k/dt) = 0$, $k = 1,\ldots,n$, we obtain (Chen and Zhang 2004)

$$E\frac{dV(\bar{X}(c,t))}{dt} = E\left(\frac{\partial V(\bar{X}(c,t))}{\partial \bar{X}(c,t)}\right)^T \bar{f}(\bar{X},c,r,t) + E\left(\frac{\partial V(\bar{X}(c,t))}{\partial \bar{X}(c,t)}\right)^T \bar{v}(t)$$

$$+\frac{1}{2}E\sum_{k=1}^{n}\bar{g}_k^T(\bar{X},c_k,r_k,t)\bar{M}_k^T \frac{\partial^2 V(\bar{X}(c,t))}{\partial \bar{X}^2(c,t)}\bar{M}_k\bar{g}_k(\bar{X},c_k,r_k,t) \tag{6.42}$$

Substituting (6.42) into (6.41), we get

$$E \int_0^{t_f} \bar{Y}^T(c,t) \bar{Q} \bar{Y}(c,t) dt = EV(\bar{X}(c,0)) - EV(\bar{X}(c,t_f))$$

$$+ E \int_0^{t_f} \left[\bar{X}^T(c,t) \bar{H}^T \bar{Q} \bar{H} \bar{X}(c,t) + \left(\frac{\partial V(\bar{X}(c,t))}{\partial \bar{X}(c,t)} \right)^T \bar{f}(\bar{X},c,r,t) + \left(\frac{\partial V(\bar{X}(c,t))}{\partial \bar{X}(c,t)} \right)^T \bar{v}(t) \right.$$

$$\left. + \frac{1}{2} \sum_{k=1}^{n} \bar{g}_k^T(\bar{X},c_k,r_k,t) \bar{M}_k^T \frac{\partial^2 V(\bar{X}(c,t))}{\partial \bar{X}^2(c,t)} \bar{M}_k \bar{g}_k(\bar{X},c_k,r_k,t) \right] dt \tag{6.43}$$

By the fact that for any vector with the desired noise attenuation level $\rho_d a^T b + b^T a \leq \rho_d^{-2} a^T a + \rho_d^2 b^T b$, (6.43) is derived as

$$E \int_0^{t_f} \bar{Y}^T(c,t) \bar{Q} \bar{Y}(c,t) dt \leq EV(\bar{X}(c,0)) + E \int_0^{t_f} \left[\bar{X}^T(c,t) \bar{H}^T \bar{Q} \bar{H} \bar{X}(c,t) \right.$$

$$+ \left(\frac{\partial V(\bar{X}(c,t))}{\partial \bar{X}} \right)^T \bar{f}(\bar{X},c,r,t) + \frac{1}{4\rho_d^2} \left(\frac{\partial V(\bar{X}(c,t))}{\partial \bar{X}} \right)^T \left(\frac{\partial V(\bar{X}(c,t))}{\partial \bar{X}} \right)$$

$$\left. + \frac{1}{2} \sum_{k=1}^{n} \bar{g}_k^T(\bar{X},c_k,r_k,t) \bar{M}_k^T \frac{\partial^2 V(\bar{X}(c,t))}{\partial \bar{x}^2} \bar{M}_k \bar{g}_k(\bar{X},c_k,r_k,t) + \rho_d^2 \bar{v}^T(t) \bar{v}(t) \right] dt \tag{6.44}$$

If the inequality in (6.26) holds, then we have

$$E \int_0^{t_f} \bar{Y}^T(c,t) \bar{Q} \bar{Y}(c,t) \, dt \leq EV(\bar{X}(c,0)) + E \int_0^{t_f} \rho_d^2 \bar{v}(t) \bar{v}(t) \, dt \tag{6.45}$$

where ρ_d is the desired noise attenuation level. If the initial $\bar{X}(c,0) = 0$, then (6.45) will be reduced to the H_∞ noise attenuation level ρ_d in (6.24). Hence the H_∞ noise attenuation level ρ_d is guaranteed if the inequality in (6.26) holds.

Appendix 6.2: Proof of Proposition 6.2

Let us choose a Lyapunov function $V(\bar{X}(c,t)) > 0$ for $\bar{X}(c,t) \neq 0$ with $V(0) = 0$ for the synthetic gene network in (6.23). Following (6.41), (6.42) and by the fact

$$\left(\frac{\partial V(\bar{X}(c,t))}{\partial \bar{X}(c,t)} \right)^T \bar{v}(t) \leq \frac{1}{4} \left(\frac{\partial V(\bar{X}(c,t))}{\partial \bar{X}(c,t)} \right)^T \left(\frac{\partial V(\bar{X}(c,t))}{\partial \bar{X}(c,t)} \right) + \bar{v}^T(t) \bar{v}(t),$$

we get

$$E \int_0^{t_f} \bar{Y}^T(c,t) \bar{Q} \bar{Y}(c,t) dt \leq EV(\bar{X}(c,0)) + E \int_0^{t_f} \left[\bar{X}^T(c,t) \bar{H}^T \bar{Q} \bar{H} \bar{X}(c,t) \right.$$

$$+ \left(\frac{\partial V(\bar{X}(c,t))}{\partial \bar{X}} \right)^T \bar{f}(\bar{X},c,r,t) + \frac{1}{4} \left(\frac{\partial V(\bar{X}(c,t))}{\partial \bar{X}} \right)^T \left(\frac{\partial V(\bar{X}(c,t))}{\partial X} \right)$$

$$\left. + \frac{1}{2} \sum_{k=1}^{n} \bar{g}_k^T(\bar{X},c_k,r_k,t) \bar{M}_k^T \frac{\partial^2 V(\bar{X}(c,t))}{\partial \bar{x}^2} \bar{M}_k \bar{g}_k(\bar{X},c_k,r_k,t) + \bar{v}^T(t) \bar{v}(t) \right] dt \tag{6.46}$$

Therefore, if the HJI in (6.27) holds, we have

$$E \int_0^{t_f} \bar{Y}^T(c,t) \bar{Q} \bar{Y}(c,t) \, dt \leq EV(\bar{X}(c,0)) + E \int_0^{t_f} \bar{v}^T(t) \bar{v}(t) \, dt \tag{6.47}$$

In other words, $EV\left(\overline{X}(c,0)\right)+E\int_{0}^{t_f}\overline{v}^{T}(t)\overline{v}(t)\,dt$ is the upper bound of the tracking error energy $E\int_{0}^{t_f}\overline{Y}(c,t)\overline{Q}\overline{Y}(c,t)\,dt$. Since the suboptimal tracking design is to minimize its upper bound instead of the tracking error energy, the suboptimal tracking design for a synthetic gene network is to select a promoter set to minimize $EV\left(\overline{X}(c,0)\right)+E\int_{0}^{t_f}\overline{v}^{T}(t)\overline{v}(t)\,dt$. Since the environmental disturbance term $E\int_{0}^{t_f}\overline{v}^{T}(t)\overline{v}(t)\,dt$ is independent on the choice of promoter set. Therefore the suboptimal tracking design of a synthetic gene network is reduced to $\min\limits_{\substack{c\in Lib_j \\ j=1,\ldots,m}} EV\left(\overline{X}(c,0)\right)$ subject to HJI in (6.27).

Appendix 6.3: Proof of Proposition 6.3

Let us denote a Lyapunov energy function $V\left(\overline{X}(c,t)\right)=\overline{X}^{T}(c,t)P\overline{X}(c,t)>0$ for $\overline{X}(c,0)\neq 0$ with $V(0)=0$. By using the T-S fuzzy approximation method, we have

$$\overline{f}(\overline{X},c,r,t)=\sum_{i=1}^{L}h_{i}(X_{i}(c,t))\overline{A}_{i}(c)\overline{X}(c,t) \tag{6.48}$$

and

$$\overline{g}_{k}\left(\overline{X},c_{k},r_{k},t\right)=\sum_{i=1}^{L}h_{i}(X_{i}(c,t))\overline{B}_{i}(c)\overline{X}(c,t) \tag{6.49}$$

Substituting (6.48) and (6.49) into HJIs in (6.26) and (6.27), we get

$$\sum_{i=1}^{L}h_{i}\left(X_{i}(c,t)\right)\left\{\overline{X}^{T}(c,t)\overline{A}_{i}^{T}(c)P\overline{X}(c,t)+\overline{X}^{T}(c,t)P\overline{A}_{i}(c)\overline{X}(c,t)\right.$$

$$\left.+\sum_{k=1}^{n}\overline{X}^{T}(c,t)\overline{B}_{i}^{T}(c)\overline{M}_{k}^{T}P\overline{M}_{k}\overline{B}_{i}(c)\overline{X}(c,t)+\frac{1}{\rho_{d}^{2}}\overline{X}^{T}(c,t)PP\overline{X}(c,t)\right.$$

$$\left.+\overline{X}^{T}(c,t)\overline{H}^{T}\overline{Q}\overline{H}\overline{X}(c,t)\right\}<0 \tag{6.50}$$

and

$$\sum_{i=1}^{L}h_{i}\left(X_{i}(c,t)\right)\left\{\overline{X}^{T}(c,t)\overline{A}_{i}^{T}(c)P\overline{X}(c,t)+\overline{X}^{T}(c,t)P\overline{A}_{i}(c)\overline{X}(c,t)\right.$$

$$\left.+\sum_{k=1}^{n}\overline{X}^{T}(c,t)\overline{B}_{i}^{T}(c)\overline{M}_{k}^{T}P\overline{M}_{k}\overline{B}_{i}(c)\overline{X}(c,t)+\overline{X}^{T}(c,t)PP\overline{X}(c,t)\right.$$

$$\left.+\overline{X}^{T}(c,t)\overline{H}^{T}\overline{Q}\overline{H}\overline{X}(c,t)\right\}<0 \tag{6.51}$$

or equivalently

$$\overline{A}_{i}^{T}(c)P+P\overline{A}_{i}(c)+\sum_{k=1}^{n}\overline{B}_{i}^{T}(c)\overline{M}_{k}^{T}P\overline{M}_{k}\overline{B}_{i}(c)+\frac{1}{\rho_{d}^{2}}PP+\overline{H}^{T}\overline{Q}\overline{H}<0 \tag{6.52}$$

and

$$\overline{A}_i^T(c)P + P\overline{A}_i(c) + \sum_{k=1}^{n} \overline{B}_i^T(c)\overline{M}_k^T P\overline{M}_k \overline{B}_i(c) + PP + \overline{H}^T\overline{Q}\overline{H} < 0 \tag{6.53}$$

By Schur complements (Boyd et al. 1994), (6.52) and (6.53) are equivalent to the LMIs in (6.32). Finally,

$$E\{V(\overline{X}(c,0))\} = E\{\overline{X}^T(c,0)P\overline{X}(c,0)\} = Tr\left(E\{P\overline{X}(c,0)\overline{X}^T(c,0)\}\right) = TrPR_0 \tag{6.54}$$

where $R_0 = E\{\overline{X}(c,0)\overline{X}^T(c,0)\}$.

Therefore, the HJI-constrained optimization in (6.28) is equivalent to the LMIs-constrained optimization problem in (6.32).

References

Alon, U. 2007. An Introduction to Systems Biology: Design Principles of Biological Circuits. Chapman & Hall/CRC, London.

Basu, S., Gerdman, Y., Collins, C.H., Arnold, F.H. and Weiss, R. 2005. A synthetic multicellular system for programmed pattern formation. Nature 434: 1130–1134.

Boyd, S., El Ghaoui, L., Feron, E. and Balakrishnan, V. 1994. Linear Matrix Inequalities in System and Control Theory. Society for Industrial Mathematics, Philadelphia.

Chen, B.S., Tseng, C.S. and Uang, H.J. 2000. Mixed H-2/H-infinity fuzzy output feedback control design for nonlinear dynamic systems: An LMI approach. IEEE Transactions on Fuzzy Systems 8: 249–265.

Chen, B.S. and Zhang, W. 2004. Stochastic H-2/H-infinity control with state-dependent noise. IEEE Transactions on Automatic Control 49: 45–57.

Chen, B.S. and Wu, C.H. 2009. A systematic design method for robust synthetic biology to satisfy design specifications. BMC Syst Biol 3: 66.

Chen, G. and Hsu, S.-H. 1995. Linear Stochastic Control Systems. Boca Raton, FL: CRC Press.

Elowitz, M.B. and Leibler, S. 2000. A synthetic oscillatory network of transcriptional regulators. Nature 403: 335–338.

Gardner, T.S., Cantor, C.R. and Collins, J.J. 2000. Construction of a genetic toggle switch in *Escherichia coli*. Nature 403: 339–342.

Goldberg, D.E. 1989. Genetic algorithms in search, optimization, and machine learning. Addison-Wesley Pub. Co., Reading, Mass.

Grefenstette, J.J. 1986. Optimization of Control Parameters for Genetic Algorithms. IEEE Transactions on Systems Man and Cybernetics 16: 122–128.

Hooshangi, S., Thiberge, S. and Weiss, R. 2005. Ultra sensitivity and noise propagation in a synthetic transcriptional cascade. Proc Natl Acad Sci USA 102: 3581–3586

Kobayashi, T., Chen, L. and Aihara, K. 2003. Modeling genetic switches with positive feedback loops. J Theor Biol 221: 379–399.

Lu, T.K., Khalil, A.S. and Collins, J.J. 2009. Next-generation synthetic gene networks. Nat Biotechnol 27: 1139–1150.

Purnick, P.E. and Weiss, R. 2009. The second wave of synthetic biology: from modules to systems. Nat Rev Mol Cell Biol 10: 410–422.

Takagi, T. and Sugeno, M. 1985. Fuzzy Identification of Systems and Its Applications to Modeling and Control. IEEE Transactions on Systems Man and Cybernetics 15: 116–132.

Wu, C.H., Lee, H.C. and Chen, B.S. 2011a. Robust synthetic gene network design via library-based search method. Bioinformatics 27: 2700–2706.
Wu, C.H., Zhang, W. and Chen, B.S. 2011b. Multiobjective H2/Hinfinity synthetic gene network design based on promoter libraries. Math Biosci 233: 111–125.
Zhang, W. and Chen, B.S. 2006. State feedback H∞ control for a class of nonlinear stochastic systems. SIAM Journal on Control and Optimization 44: 1973–1991.

7

Robust Design of Synthetic Biological Filter and Transistor based on Promoter-RBS Libraries

In this chapter, two design examples, namely, robust design of synthetic biological filter and transistor, are given to demonstrate the systematic design methodology of synthetic gene networks based on promoter-RBS libraries constructed in Chapter 5.

7.1 Systematic Design Methodology for Synthetic Biological Filter

7.1.1 A cascade gene circuit topology for biological filter design

After promoter-RBS libraries are built from the promoter-RBS strengths, a set of biological components from the promoter-RBS libraries in Table 5.4–5.7 and the corresponding regulatory protein genes are selected to engineer a biological filter on the basis of the cascade gene circuit topology in Figure 7.1. The synthetic gene circuit functions as a biological filter with inducers I_1 and I_2 that induce the activities of corresponding promoter-RBS components, repressor-regulated promoter-RBS component R_i or activator-regulated promoter-RBS component A_i. When one of the inducers is taken as input to the synthetic cascade gene circuit, the threshold (cutoff) of the biological input/output (I/O) filtering response can be tuned externally by changing the concentration of another inducer in Figure 7.1. The threshold is the point that must be exceeded to begin producing a given effect or to elicit a response.

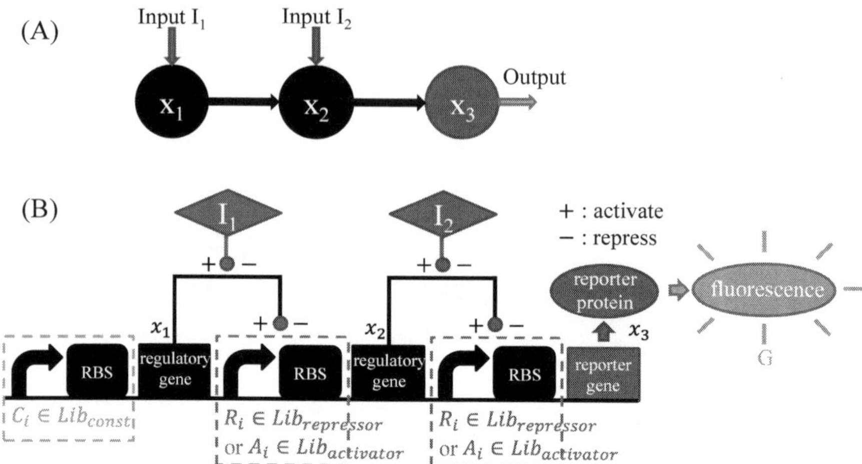

$C_i \in Lib_{const}$

$R_i \in Lib_{repressor}$ or $A_i \in Lib_{activator}$

$R_i \in Lib_{repressor}$ or $A_i \in Lib_{activator}$

Figure 7.1. (A) A synthetic biological filter design based on a cascade gene circuit topology. (B) A synthetic cascade circuit with three promoter-RBS components is designed to function as a biological filter. The first node has been fixed by a constitutive promoter-RBS component. The second and third nodes can be engineered by an activator-regulated promoter-RBS component or a repressor-regulated promoter-RBS component, respectively. When different kinds of promoter-RBS components and regulatory genes are selected, they activate (+ sign) or repress (− sign) the connected node.

Color image of this figure appears in the color plate section at the end of the book.

For the convenience of description and explanation, a synthetic biological filter circuit is assembled by selecting a set of biological components and genes, namely, a constitutive promoter-RBS component C_{1i}, two different kinds of repressor-regulated promoter-RBS components R_{2i} and R_{3i}, two regulated repressor genes, and an appropriate reporter gene, as shown in Figure 7.2A. First, if the inducer I_1 and the fluorescence expression of the reporter protein are taken as input and output, respectively, the low-pass I/O filtering response can be obtained. The biological low-pass filter would be maintained at high expression when input concentration I_1 is low, and would be maintained at low expression when input concentration I_1 is high, as shown in Figure 7.2B. In the proposed method, an additional design feature is involved: specifically, the inducer concentration I_2 in Figure 7.2A can be adjusted to precisely and externally tune the threshold of the biological low-pass filter. In the biological filter design, the threshold is the point at which the system I/O response transfers from high to low or from low to high in Figure 7.2B and C, respectively. In addition to the external control of inducer concentration I_2, promoter-RBS components with different strengths can be selected to enable distinct thresholds based on promoter-RBS libraries. This flexible design feature is novel in synthetic biological filters.

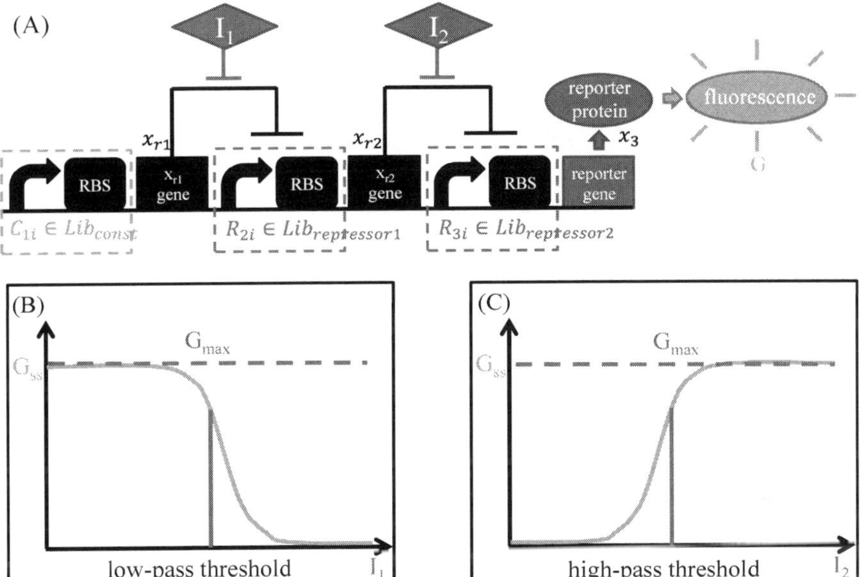

Figure 7.2. A synthetic biological filter with promoter-RBS components selected from the constitutive and repressor-regulated promoter-RBS libraries, and the I/O response of the biological low-pass and high-pass filters. (A) For convenience of explaining the operating mechanism of the biological filter, a synthetic biological filter is assembled by selecting a constitutive promoter-RBS component C_{1i}, and two different repressor-regulated promoter-RBS components R_{2i} and R_{3i} from their corresponding libraries, to regulate their corresponding regulatory genes and reporter gene. (B) The I/O response of the low-pass filter for input inducer I_1 and output fluorescence G_{ss} at steady state. (C) The I/O response of the high-pass filter for input inducer I_2 and output fluorescence G_{ss} at steady state.

Color image of this figure appears in the color plate section at the end of the book.

To illustrate the operating mechanism of the biological low-pass filter in Figure 7.2A, the reporter gene is under the control of repressor x_{r2} which is repressed by repressor x_{r1}. When the inducer I_1 is absent, a continuous production of repressor x_{r1} completely inhibits the generation of repressor x_{r2}. Therefore, the fluorescence expression without inhibition would remain at a high level. When the input inducer concentration I_1 is gradually increased, the repressor activity of repressor x_{r1} is reduced because inducer I_1 binds to repressor x_{r1} to prevent free repressor x_{r1} from binding with the corresponding promoter-RBS component R_{2i}. As a result, repressor x_{r2} is generated, thus inhibiting the production of reporter protein. Therefore, fluorescence is inhibited and converted from a high level to a low level, with a low-pass I/O filtering response as shown in Figure 7.2B. Notably, the concentration of another inducer I_2 functions as a buffer to adjust the I/O response of the biological low-pass filter to the input inducer I_1.

Fluorescence expression is not inhibited by low concentrations of repressor x_{r2}, since inducer I_2 binds to repressor x_{r2} to prevent free repressor x_{r2} from binding to the corresponding promoter-RBS component R_{3i}. Hence, the threshold for the biological low-pass filter is pushed forward when the inducer concentration I_1 is increased gradually. This is why the threshold of the biological low-pass filter can be precisely tuned by changing the control inducer concentration I_2 externally.

The operation mechanism of the biological high-pass filter is also shown in Figure 7.2A, where fluorescence expression of the reporter gene is under the control of the repressor x_{r2}. When the inducer I_2 is absent, the repressor x_{r2} completely binds to the corresponding promoter-RBS component R_{3i}. Therefore, the production of immature reporter protein and fluorescence expression would be inhibited. When the inducer concentration I_2 increases, the repressor activity of repressor x_{r2} decreases because inducer I_2 binds to repressor x_{r2} to prevent free repressor x_{r2} from binding to the corresponding promoter-RBS component R_{3i}. Thus, fluorescence expression is no longer inhibited and is converted from a low level to a high level, with a high-pass filtering response as shown in Figure 7.2C. Notably, the inducer concentration I_1 functions as a regulatory mechanism to regulate the production of repressor concentration x_{r2} of the biological high-pass filter. The inducer concentration I_1 regulates the gene expression of repressor x_{r2} because I_1 binds to the repressor x_{r1}. Regulation is driven by the constitutive promoter-RBS component C_{1i}, which prevents the free repressor x_{r1} from binding to the corresponding promoter-RBS component R_{2i}. When the concentration of inducer I_1 increases, the production of repressor x_{r2} then increases, thus inhibiting fluorescence expression. Therefore, the threshold for the biological high-pass filter would be pushed forward when the inducer concentration I_1 is increased gradually. This is why the threshold of the biological high-pass filter can be precisely tuned by changing the control inducer concentration of I_1 externally.

7.1.2 Dynamic model of the biological filter based on gene circuit topology

The cascade gene circuit topology is depicted in Figure 7.1, from which the dynamic model of the synthetic biological filter can be introduced. The concentrations of two different regulatory proteins are denoted by x_1 and x_2, and the concentrations of the immature and mature reporter proteins are denoted by x_3 and G, respectively. In the following discussion, the concentration of mature reporter protein is assumed to be equal to that measured by ELISA reader. The dynamic model of the biological filter is given as follows:

$$\begin{cases} \dot{x}_1(t) = P(S_{1u}, S_{1l}, 0, 0) - (\mu + \gamma_{x_1})x_1(t) + \omega_1(t) \\ \dot{x}_2(t) = P(S_{2u}, S_{2l}, x_1, I_1) - (\mu + \gamma_{x_2})x_2(t) + \omega_2(t) \\ \dot{x}_3(t) = P(S_{3u}, S_{3l}, x_2, I_2) - (m + \mu + \gamma_{x_3})x_3(t) + \omega_3(t) \\ \dot{G}(t) = mx_3 - (\mu + \gamma_G)G(t) + \omega_4(t) \end{cases} \tag{7.1}$$

where γ_{x_1} and γ_{x_2} are the degradation rates of two different regulatory proteins x_1 and x_2; γ_{x_3} and γ_G are the degradation rates of the immature and mature reporter proteins, respectively; $\omega_1(t)$, $\omega_2(t)$ and $\omega_3(t)$ denote the molecular noises in the transcription and translation processes; $\omega_4(t)$ denotes the molecular noise in mature protein expression; $\{S_{1u}, S_{1l}\}$, $\{S_{2u}, S_{2l}\}$ and $\{S_{3u}, S_{3l}\}$ denote the maximum and minimum strengths of three different promoter-RBS components from the corresponding libraries; I_1 and I_2 denote the inducer concentrations for regulating the activities of x_1 and x_2, respectively. On the basis of actual synthetic gene circuit, the promoter-RBS regulation function $P(S_u, S_l, TF, I)$ with maximum and minimum promoter-RBS strengths S_u and S_l, respectively, the corresponding transcription factor concentration TF and related inducer concentration I in (7.1) can be substituted with the constitutive promoter-RBS regulation function in (5.10) for the constitutive promoter-RBS component C_i. Similarly, it can be substituted with the repressor-regulated promoter-RBS regulation function in (5.12) for repressor-regulated promoter-RBS component R_i, or with the activator-regulated promoter-RBS regulation function in (5.14) for the activator-regulated promoter-RBS component A_i. The transcription factor concentration may be equal to zero for the constitutive promoter-RBS component C_i. Alternatively, it may be equal to the repressor concentration x_r for the repressor-regulated promoter-RBS component R_i, and activator concentration x_a for the activator-regulated promoter-RBS component A_i.

For the convenience of design, a synthetic biological filter is assembled by selecting a set of specified biological components, i.e., a constitutive promoter-RBS component C_{1i}, two different kinds of repressor-regulated promoter-RBS components R_{2i} and R_{3i}, two regulated repressor genes, and an appropriate reporter gene, as shown in Figure 7.2A. In this case, two different regulatory protein concentrations x_1 and x_2 in (7.1) are replaced by two different repressor concentrations denoted by x_{r1} and x_{r2}, respectively. The first promoter-RBS regulation function $P(S_{1u}, S_{1l}, 0, 0)$ in (7.1) is substituted with the constitutive promoter-RBS regulation function in (5.10). The second and third promoter-RBS regulation functions $P(S_{2u}, S_{2l}, x_1, I_1)$ and $P(S_{3u}, S_{3l}, x_2, I_2)$, respectively, in (7.1) are substituted with the repressor-regulated promoter-RBS regulation function in (5.12). On the basis of the gene circuit topology in Figure 7.2A, the dynamic model for the biological filter in (7.1), which has a set of promoter-RBS components C_{1i}, R_{2i}, and R_{3i} selected from the promoter-RBS libraries Lib_{const}, $Lib_{repressor1}$, and $Lib_{repressor2}$, respectively, is rewritten as follows

$$\begin{cases} \dot{x}_{r1}(t) = P_c(S_{1u,i},0,0,0) - (\mu + \gamma_{x_{r1}})x_{r1}(t) + \omega_1(t) \\ \dot{x}_{r2}(t) = P_r(S_{2u,i},S_{2l,i},x_{r1},I_1) - (\mu + \gamma_{x_{r2}})x_{r2}(t) + \omega_2(t) \\ \dot{x}_3(t) = P_r(S_{3u,i},S_{3l,i},x_{r2},I_2) - (m + \mu + \gamma_{x_3})x_3(t) + \omega_3(t) \\ \dot{G}(t) = mx_3(t) - (\mu + \gamma_G)G(t) + \omega_4(t) \end{cases}$$

$$\{S_{1u,i}\} \in Lib_{const}, \{S_{2u,i},S_{2l,i}\} \in Lib_{repressor1}, \{S_{3u,i},S_{3l,i}\} \in Lib_{repressor2}$$

where $P_c(S_{1u,i},0,0,0) = S_{1u,i}$

$$P_r(S_{2u,i},S_{2l,i},x_{r1},I_1) = S_{2l,i} + \frac{S_{2u,i}-S_{2l,i}}{1+\left(\dfrac{x_{r1}^*(x_{r1},I_1)}{K_{r2}}\right)^{n_{r2}}}, x_{r1}^*(x_{r1},I_1) = \frac{x_{r1}}{1+\dfrac{I_1}{K_{I_1}}} \qquad (7.2)$$

$$P_r(S_{3u,i},S_{3l,i},x_{r2},I_2) = S_{3l,i} + \frac{S_{3u,i}-S_{3l,i}}{1+\left(\dfrac{x_{r2}^*(x_{r2},I_2)}{K_{r3}}\right)^{n_{r3}}}, x_{r2}^*(x_{r2},I_2) = \frac{x_{r2}}{1+\dfrac{I_2}{K_{I_2}}}$$

where $\gamma_{x_{r1}}$ and $\gamma_{x_{r2}}$ are the degradation rates of repressors x_{r1} and x_{r2}, respectively; $\{S_{1u,i}\} \in Lib_{const}$ denotes the promoter-RBS strength $S_{1u,i}$ of the constitutive promoter-RBS component C_{1i}, which is selected from Lib_{const}. $\{S_{2u,i},S_{2l,i}\} \in Lib_{repressor1}$ denotes the maximum and minimum promoter-RBS strengths $S_{2u,i}$ and $S_{2l,i}$, respectively, of the repressor-regulated promoter-RBS components R_{2i} selected from $Lib_{repressor1}$. $\{S_{3u,i},S_{3l,i}\} \in Lib_{repressor2}$ denotes the maximum and minimum promoter-RBS strengths $S_{3u,i}$ and $S_{3l,i}$, respectively, of the repressor-regulated promoter-RBS components R_{3i} selected from $Lib_{repressor2}$. K_{r2} and n_{r2} denote the binding affinity and the binding cooperativity between repressor x_{r1} and promoter-RBS component R_{2i}, respectively, from the promoter-RBS library $Lib_{repressor1}$. K_{r3} and n_{r3} defined in the corresponding promoter-RBS library $Lib_{repressor2}$ denote the binding affinity and the binding cooperativity between repressor x_{r2} and promoter-RBS component R_{3i}, respectively. K_{I_1} and K_{I_2} denote the dissociation rate between inducers I_1, I_2 and repressors x_{r1}, x_{r2}, respectively. Therefore, the concentration of inducers I_1 and I_2 affected the repressor activities $x_{r1}^*(x_{r1},I_1)$ and $x_{r2}^*(x_{r2},I_2)$, respectively, thereby affecting the strengths of promoter regulation functions $P_r(S_{2u,i},S_{2l,i},x_{r1},I_1)$ and $P_r(S_{3u,i},S_{3l,i},x_{r2},I_2)$ in (7.2), respectively. Eventually, the expression of repressor x_{r2} and reporter protein GFP is affected, as shown in Figure

7.2A. Meanwhile, the expression of repressor x_{r1} is not affected by the concentrations of inducers I_1 and I_2 because the repressor x_{r1} is driven by a constitutive promoter-RBS component C_{1i} without external regulation.

For the I/O response of the biological filter, the dynamic model is transformed to a steady-state model by assuming that the system dynamics in (7.2) are equal to zero. The steady-state concentrations of two repressors, as well as immature and mature reporter proteins are denoted by x_{r1ss}, x_{r2ss}, x_{3ss} and G_{ss}, respectively. The steady-state model of the biological filter system with promoters C_{1i}, R_{2i} and R_{3i} selected from the corresponding promoter-RBS libraries Lib_{const}, $Lib_{repressor1}$, and $Lib_{repressor2}$ respectively, is derived as follows:

$$
\begin{cases}
x_{r1ss} = \dfrac{P_c(S_{1u,i},0,0,0)}{\mu + \gamma_{x_{r1}}} + \omega_1 \\[2ex]
x_{r2ss} = \dfrac{P_r(S_{2u,i},S_{2l,i},x_{r1},I_1)}{\mu + \gamma_{x_{r2}}} + \omega_2 \\[2ex]
x_{3ss} = \dfrac{P_r(S_{3u,i},S_{3l,i},x_{r2},I_2)}{m + \mu + \gamma_{x_3}} + \omega_3 \\[2ex]
G_{ss} = \dfrac{m}{\mu + \gamma_G}x_{3ss} + \omega_4
\end{cases}
\qquad (7.3)
$$

$$\{S_{1u,i}\} \in Lib_{const}, \quad \{S_{2u,i},S_{2l,i}\} \in Lib_{repressor1}, \quad \{S_{3u,i},S_{3l,i}\} \in Lib_{repressor2}$$

where ω_1, ω_2 and ω_3 denote the molecular noises in the transcription and translation processes of gene expression at steady-state; ω_4 denotes the molecular noises in mature protein expression. The steady-state protein expression of the biological filter system can be obtained from the steady-state model (7.3) with the promoter regulation functions $P_c(S_{1u,i},0,0,0)$, $P_r(S_{2u,i},S_{2l,i},x_{r1},I_1)$, and $P_r(S_{3u,i},S_{3l,i},x_{r2},I_2)$; and repressor activities $x_{r1}^*(x_{r1},I_1)$ and $x_{r2}^*(x_{r2},I_2)$ represented in (7.2).

Next, the biological low-pass I/O filtering response is analyzed on the basis of the steady-state model (7.3) of gene circuit topology in Figure 7.2A. When the inducer I_1 is absent, the repressor activity $x_{r1}^*(x_{r1},I_1)$ is equal to the repressor concentration x_{r1}. Since the steady-state repressor concentration x_{r1ss} regulated by the constitutive promoter-RBS component C_{1i} is a constant in the steady-state model (7.3), the promoter-RBS regulation function $P_r(S_{2u,i},S_{2l,i},x_{r1},I_1)$ is close to $S_{2l,i}$, the minimum promoter-RBS strength of the corresponding repressor-regulated promoter-RBS component R_{2i} in the library $Lib_{repressor1}$. The repressor activity $x_{r2}^*(x_{r2},I_2)$ is close to zero because the steady-state concentration of repressor x_{r2} regulated by repressor-regulated promoter-RBS component R_{2i} is maintained at a low level. Thus, the promoter-RBS regulation function $P_r(S_{3u,i}, S_{3l,i}, x_{r2}, I_2)$ is close to $S_{3u,i}$, the maximum promoter-RBS strength. Finally, the steady-state concentration of immature reporter protein x_{3ss} is maintained at a high level and the steady-state GFP fluorescence expression G_{ss} in (7.3) is also maintained at a high level. As the inducer concentration I_1 gradually increases, the repressor activity $x_{r1}^*(x_{r1}, I_1)$ decreases. The promoter-RBS regulation function $P_r(S_{2u,i}, S_{2l,i}, x_{r1}, I_1)$ increases to a level greater than $S_{2l,i}$, and the steady-state concentration of repressor x_{r2} (7.3) in gradually increases. Thus, the repressor activity $x_{r2}^*(x_{r2}, I_2)$ is increased at a fixed concentration of inducer I_2, and the value of the promoter-RBS regulation function $P_r(S_{3u,i}, S_{3l,i}, x_{r2}, I_2)$ decreases to a level lower than $S_{3u,i}$. Therefore, the steady-state concentration of the immature reporter protein x_{3ss} gradually decreases, and the steady-state fluorescence expression G_{ss} in (7.3) also decreases. The repressor activity $x_{r2}^*(x_{r2}, I_2)$ can be regulated by the inducer concentration I_2. For example, a higher repressor concentration x_{r2} regulated by the inducer I_1 is needed to achieve the same repressor activity $x_{r2}^*(x_{r2}, I_2)$ when the inducer concentration I_2 is increased. This is why the threshold of the biological low-pass filter can be precisely tuned by the concentration of external inducer I_2.

Meanwhile, the biological high-pass I/O filtering response is analyzed based on the steady-state model (7.3) of gene circuit topology in Figure 7.2A. When the inducer I_2 is absent, repressor activity $x_{r2}^*(x_{r2}, I_2)$ is equal to the repressor concentration x_{r2}. The promoter-RBS regulation function $P_r(S_{3u,i}, S_{3l,i}, x_{r2}, I_2)$ would be close to $S_{3l,i}$, the minimum promoter-RBS strength of the promoter-RBS component R_{3i} in the library $Lib_{repressor\,2}$. Hence, the steady-state concentration of immature reporter protein x_{3ss} and fluorescence G_{ss} would be maintained at a low level in (7.3). As the inducer concentration I_2 is gradually increased, the repressor activity $x_{r2}^*(x_{r2}, I_2)$ is decreased. The promoter-RBS regulation function $P_r(S_{3u,i}, S_{3l,i}, x_{r2}, I_2)$ would be increased to a level greater than $S_{3l,i}$. Therefore, the steady-state concentration of immature reporter protein x_{3ss} and steady-state fluorescence expression G_{ss} would gradually increase until the promoter-RBS regulation function $P_r(S_{3u,i}, S_{3l,i}, x_{r2}, I_2)$ reaches the maximum strength $S_{3u,i}$. The repressor activity $x_{r1}^*(x_{r1}, I_1)$ can be regulated by the concentration of inducer I_1. The repressor activity $x_{r1}^*(x_{r1}, I_1)$ would then influence the promoter-RBS regulation function $P_r(S_{2u,i}, S_{2l,i}, x_{r1}, I_1)$, thereby affecting the steady state concentration of repressor x_{r2} in (7.3). For example, when the concentration of inducer I_1 is increased, the expression of repressor x_{r2} is increased. To achieve the same repressor activity $x_{r2}^*(x_{r2}, I_2)$, a higher inducer concentration I_2 is needed to bind increased repressor concentration x_{r2}. This is why the threshold of the biological high-pass filter can be precisely tuned by the concentration of external inducer I_1.

7.1.3 Formulation of design specifications for biological filters

Our design purpose is to engineer a synthetic biological filter by selecting an adequate set of promoter-RBS components from the corresponding libraries and an appropriate control inducer concentration to achieve the optimal reference matching of the desired I/O response. When the steady-state model of the biological filter is represented in (7.3), a desired I/O response of fluorescence expression with different input concentrations needs to be prescribed for matching of biological filtering performance. The desired I/O responses for the biological low-pass and high-pass filters denoted by $G_{r1}(u)$ and $G_{r2}(u)$, respectively, can be specified as follows:

$$G_{r1}(u) = \frac{1}{\sqrt{1^2 + (\frac{u}{u_{lc}})^2}} G_{\max} \tag{7.4}$$

$$G_{r2}(u) = \frac{\frac{u}{u_{hc}}}{\sqrt{1^2 + (\frac{u}{u_{hc}})^2}} G_{\max} \tag{7.5}$$

where $G_{\max}$ denotes the maximum fluorescence expression of the biological filter with an adequate set of promoter-RBS components selected from the corresponding promoter-RBS libraries; u is the input signal to the filter system. For a desired biological low-pass I/O response in (7.4), the input signal u is the concentration of inducer I_1 based on the biological filter circuit in Figure 7.2A. For a desired biological high-pass response in (7.5), the input signal u is the concentration of inducer I_2 based on the biological filter circuit in Figure 7.2A. u_{lc} denotes the threshold specified for the biological low-pass filtering response and u_{hc} denotes the threshold specified for the biological high-pass filtering response. For the desired I/O response of the biological low-pass filter in (7.4), a low concentration of input inducer I_1, such that $I_1 << u_{lc}$, leads to high fluorescence levels close to the maximum expression; a high concentration of input inducer I_1, such that $I_1 >> u_{lc}$, leads to low fluorescence levels close to zero. On the other hand, for the desired I/O response of the biological high-pass filter in (7.5), a low concentration level of input inducer I_2, specifically, $I_2 << u_{hc}$, leads to low fluorescence levels; and a high concentration level of input inducer I_2, specifically, $I_2 >> u_{hc}$, leads to high fluorescence levels. When the input signal u is equal to the low-pass threshold u_{lc} or high-pass threshold u_{hc}, the fluorescence expression of the biological filter reach $1/\sqrt{2}$ times the maximum fluorescence level. On the basis of the gene circuit topology and the steady-state model of the biological filter, the most adequate set of promoter-RBS components and an appropriate regulatory inducer concentration can be selected from the corresponding libraries to match the specified I/O responses $G_{r1}(u)$ and $G_{r2}(u)$ in (7.4) and (7.5) for the biological low-pass and high-pass filters, respectively.

7.1.4 Biological filter design via a mixed library-based searching method

After the desired I/O response is specified as described above, a library-based search method (Wu et al. 2011) is employed to efficiently select the most adequate set of promoter-RBS components from the corresponding promoter-RBS libraries to achieve the desired I/O response for the biological filter design. Furthermore, a concentration of regulatory inducer is taken into consideration as another inducer concentration library, resulting in a mixed library-based search method. After the steady-state model of the biological filter is established in (7.3), our ultimate goal is to design a biological filter to match the specified I/O response by selecting an adequate inducer concentration to be tuned externally and an appropriate set of promoter-RBS components from the corresponding promoter-RBS libraries.

In addition, the standard deviations of stochastic parameter fluctuation need to be considered in the model as biological components are inherently uncertain *in vivo* because of gene-expression noise, cell growth, DNA mutations, evolution, and parameter estimation errors. The standard deviations of stochastic parameter fluctuation and environmental molecular disturbances can be considered as quantitative specifications of system robustness for the biological filter. Here, in order to engineer an externally tunable biological filter, a mixed library-based search method with four design specifications is stated as follows:

i) Specify a desired I/O response of the biological filter with a desired maximum fluorescence G_{max}, and a specified threshold u_{lc} for low-pass response or u_{hc} for high-pass response in (7.4) and (7.5), respectively.
ii) Provide the well-characterized promoter-RBS libraries listed in Table 5.4–5.7 and a set of inducer concentration libraries.
iii) Specify the parameter uncertainties to be tolerated by the biological filter in the host cell, namely, the standard deviations σ_i of environmental disturbances of transcription and translation processes ω_i; standard deviations $\Delta S_{u,i}$ and $\Delta S_{l,i}$ of the maximum and minimum promoter-RBS strengths $S_{u,i}$ and $S_{l,i}$, respectively; and standard deviations $\Delta \gamma_{x_i}$, $\Delta \gamma_{x_3}$,

$\Delta\gamma_G$, Δm and $\Delta\mu$ of the regulatory protein degradation rate γ_{x_i}, immature and mature reporter protein degradation rates γ_{x_3} and γ_G, maturation rate m and dilution rate μ due to cell growth, respectively.

iv) Define an explicit cost function

$$J(f_i) = E \int_0^{u_f} (G_{ss}(f_i, u) - G_{ri}(u))^2 \, du \qquad (7.6)$$

$$i = 1, 2 \text{ for low-pass and high-pass filter}$$

where f_i denotes the combination $\{pR_1, pR_2, pR_3, I\}$ of promoter-RBS set $\{pR_1, pR_2, pR_3\}$ with an inducer concentration I; $G_{ss}(f_i, u)$ and $G_{ri}(u)$ are the estimated and desired steady-state fluorescence levels under the specific combination f_i with input concentration u, respectively; pR_1, pR_2, and pR_3 represent the first, second, and third promoter-RBS components selected from three types of promoter-RBS libraries Lib_{const}, $Lib_{repressor}$, and $Lib_{activator}$ based on gene circuit topology; and I denotes an appropriate control inducer concentration from library Lib_I. If $i = 1$, the synthetic gene circuit functions as a low-pass filter and takes the inducer concentrations I_1 and I_2 as the input molecular signal and control molecular signal, respectively. On the other hand, if $i = 2$, the synthetic gene circuit functions as a high-pass filter and takes the inducer concentrations I_2 and I_1 as the input molecular signal and control molecular signal, respectively.

If the cost function in (7.6) can be minimized by selecting the most adequate set $f_i^* = \{pR_1^*, pR_2^*, pR_3^*, I^*\}$ from their corresponding libraries under the design specifications (i)–(iv), then the I/O response $G_{ss}(f_i^*, u)$ of the biological filter can be optimally matched to the specified I/O response $G_{ri}(u)$ under the intrinsic parameter fluctuations and environmental disturbances to the host cell.

Current optimal design methods for synthetic gene circuits rely on exhaustive searching, which is a time-consuming process. When the scale of libraries are gradually expanded, the combinatorial explosion of promoter-RBS components require a large running time to search for an adequate promoter-RBS set from the corresponding libraries to satisfy the four design specifications. A large amount of computation time is also wasted because the complexity of the library-based search method is augmented by mixing a library of control inducer concentrations. The regulatory inducer concentration is an external adjustable library, and therefore increases the complexity when combined with the promoter-RBS set. Here, a genetic algorithm (GA)-based library search method is introduced to efficiently select the most adequate combination f_i^* from mixed libraries, specifically, a promoter-RBS set from the corresponding promoter-RBS libraries and a regulatory inducer concentration from the inducer concentration library.

This is done to minimize $J(f_i)$ in (7.6) and to achieve the desired amplitude response under the four design specifications.

Based on the mixed library-based search method for the four design specifications, the most adequate set f_i^* of a promoter-RBS set and a regulatory inducer concentration can be selected from the mixed libraries. Despite intrinsic parameter fluctuations and environmental disturbances in design specification (iii), the biological filter can be designed with a desired I/O response in design specification (i) by minimizing the cost function $J(f_i)$ in design specification (iv). The biological filter can robustly and optimally achieve the desired I/O response by using the proposed mixed library-based search method with the GA to efficiently select the most adequate set $f_i^* = \left\{ pR_1^*, pR_2^*, pR_3^*, I^* \right\}$ to satisfy the four design specifications (i)–(iv). A design procedure for the biological filter based on promoter-RBS libraries, gene circuit topology and a mixed library-based search method using GA is described below.

7.1.5 Design procedure for the biological filter

1) Build three different types of promoter-RBS libraries listed in Table 5.4–5.7 based on the dynamic characteristics and experimental data of the synthetic gene circuit.
2) Provide a cascade gene circuit topology for the biological filter design. Subsequently, construct the dynamic model as (7.2) and obtain the steady-state model as (7.3) based on the gene circuit topology.
3) Provide user-defined design specifications (i)–(iv) for biological filter with a desired low-pass or high-pass I/O response in (7.4) or (7.5), respectively.
4) Select an initial set of promoter-RBS components from the corresponding libraries and a control inducer concentration.
5) Calculate the cost function $J(f_i)$ in (7.6) for each set f_i of promoter-RBS components and control inducer concentration in the population.
6) Create offspring set f_i using GA operators such as reproduction, crossover, and mutation.
7) Calculate the cost function of a new set f_i obtained by natural selection.
8) Stop when the design goal is achieved or an acceptable solution is obtained; otherwise create a subsequent generation by returning to step 6.

The GA operators are used to generate new populations f_i of promoter-RBS components and a control inducer concentration. After new populations are evaluated by calculating the cost function in (7.6), an improved new population of candidate solutions is formed in each genetic generation. Based on the design procedure above and the mixed library-based search method, the externally tunable biological filter can be engineered to satisfy the user-defined design specifications (i)–(iv) and achieve the desired I/O response without the need for a large number of trial-and-error experiments as used in conventional methods.

7.1.6 Synthetic biological filter design

Consider the steady-state model (7.3) of the biological filter system in Figure 7.2A. On the basis of the gene circuit topology, three promoter-RBS components C_{1i}, R_{2i} and R_{3i} are selected from the constitutive promoter-RBS library Lib_{const}, and the repressor-regulated promoter-RBS libraries $Lib_{repressor1}$ and $Lib_{repressor2}$, respectively. Thereafter, the repressor genes x_{r1} and x_{r2} are embedded upstream of the corresponding repressor-regulated promoter-RBS components. In addition, an appropriate reporter protein GFP-LVA is selected to reflect immediately the gene circuit outcome. For simulation and implementation, a set of promoter-RBS components from the promoter-RBS libraries listed in Table 5.4–5.7 and their corresponding regulatory genes from BioBrick are selected to assemble the synthetic biological filter. In Figure 7.3, an example of the biological filter circuit is shown and can be divided into three cascade modules. Each module has both a promoter-RBS component selected from the corresponding promoter-RBS libraries and a specified gene connected downstream of the promoter. In the first module, the repressor LacI-LVA is expressed continuously from the constitutive promoter-RBS component C_{1i} selected from the promoter-RBS library Lib_{const}. The repressor LacI-LVA then inhibits the expression of TetR-LVA by binding to the LacI-regulated promoter-RBS component R_{2i} selected from promoter-RBS library Lib_{LacI} in the second module. Finally, the repressor TetR-LVA driven by the LacI-regulated promoter-RBS component R_{2i} inhibits the production of GFP-LVA by binding to the TetR-regulated promoter-RBS component R_{3i} selected from promoter-RBS library Lib_{TetR} in the third module. The external inducers IPTG and ATc are used to regulate the repressor activities.

For convenience, the concentrations of the repressors LacI-LVA and TetR-LVA are denoted by x_{r1} and x_{r2}, respectively. The concentration of immature GFP-LVA is denoted by x_3. The external inducer concentrations of

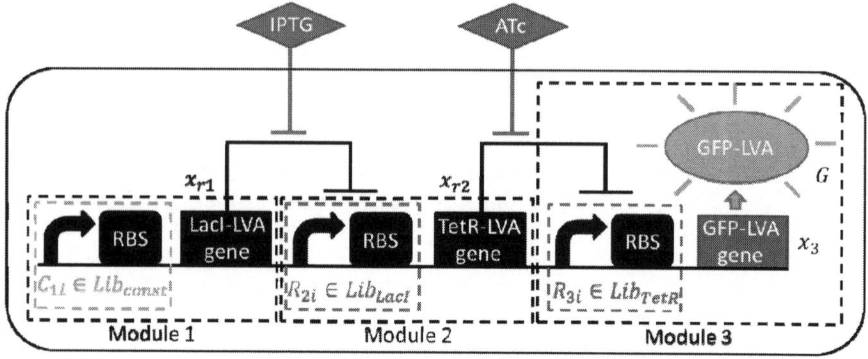

Figure 7.3. A design example of the biological low-pass filter. The biological low-pass filter can be divided into three cascade modules. Each module has both a promoter-RBS component selected from the corresponding promoter-RBS libraries and a specified gene downstream of the promoter. The first module contains a constitutive promoter-RBS component C_{1i} and a repressor LacI-LVA gene. The second module contains a LacI-regulated promoter-RBS component R_{2i} and a repressor TetR-LVA gene, and the third module contains a TetR-regulated promoter-RBS component R_{3i} and a reporter protein GFP-LVA gene.

Color image of this figure appears in the color plate section at the end of the book.

IPTG and ATc are denoted as I_1 and I_2, respectively. The stochastic dynamic model for the biological filter is then represented in the same manner as (7.2), with a set of promoter-RBS components C_{1i}, R_{2i}, and R_{3i} selected from the promoter-RBS libraries Lib_{const}, Lib_{LacI}, and Lib_{TetR}, respectively:

$$\{S_{1u,i}\} \in Lib_{const}, \{S_{2u,i}, S_{2l,i}\} \in Lib_{LacI}, \{S_{3u,i}, S_{3l,i}\} \in Lib_{TetR} \tag{7.7}$$

Subsequently, the steady-state model of the biological filter can be derived in the same manner as (7.3), where x_{r1ss}, x_{r2ss}, and x_{3ss} denote the steady-state concentrations of LacI-LVA, TetR-LVA, and immature GFP-LVA, respectively.

According to the gene circuit topology in Figure 7.3 and steady-state model in (7.3), if the inducer IPTG is taken as an input to the filter and the inducer ATc is taken as an external control, then the low-pass I/O response in Figure 7.2B is obtained. In this case, the input signal $u = I_1 =$ IPTG and control inducer $I_2 =$ ATc. Next, the design specifications for the biological low-pass filter with the desired low-pass I/O response are prescribed as follows:

i) A desired low-pass I/O response $G_{r1}(I_1)$ of the biological low-pass filter in Figure 7.2B, with a specified threshold $u_{lc} = 10^{-4}$ mM and maximum fluorescence level $G_{max} = 4.6 \times 10^4$ MEFL, is provided as follows for biological filter matching in (7.4):

$$G_{r1}(I_1) = \frac{4.6 \times 10^4}{\sqrt{1^2 + (\dfrac{I_1}{10^{-4}})^2}} \qquad (7.8)$$

ii) The index of Lib_{const} is from 1 to 27, the index of Lib_{LacI} is from 1 to 8, and the index of Lib_{TetR} is from 1 to 8. Further, the concentration of control inducer ATc ranges from 0 mM to 4.337×10^{-6} mM.

iii) The environmental disturbances and parameter fluctuations with the standard deviations to be tolerated are assumed as follows:

$$\Delta S_{1u,i} = 0.05 S_{1u,i}$$

$$(\Delta S_{2u,i}, \Delta S_{2l,i}) = (0.05 S_{2u,i}, 0.05 S_{2l,i})$$

$$(\Delta S_{3u,i}, \Delta S_{3l,i}) = (0.05 S_{3u,i}, 0.05 S_{3l,i})$$

$$\Delta \gamma_{x_{rT}} = 0.05 \gamma_{x_{rT}}, \Delta \gamma_{x_{rL}} = 0.05 \gamma_{x_{rL}} \qquad (7.9)$$

$$\Delta \gamma_{x_3} = 0.05 \gamma_{x_3}, \Delta \gamma_G = 0.05 \gamma_G$$

$$\Delta m = 0.05 m, \Delta \mu = 0.05 \mu$$

The environmental disturbances ω_i of the transcription and translation processes are independent Gaussian white noises with zero means and unit variances.

iv) The following cost function for the biological low-pass filter needs to be minimized:

$$J(f_1^*) = \min_{f_1 = \{pR_1, pR_2, pR_3, I_2\}} E \int_0^{I_{1f}} (G_{ss}(f_1, I_1) - G_{r1}(I_1))^2 dI_1 \qquad (7.10)$$

In order to efficiently solve the constrained optimal matching design problem of the biological filter via the proposed mixed library-based search method, a GA-based library search method is employed. The most adequate set $f_1^* = \{pR_1^*, pR_2^*, pR_3^*, I_2^*\} = \{C_{19}, R_{L1}, R_{T1}, ATc = 2.125 \times 10^{-6} \text{ mM}\}$ for the biological low-pass filter is then searched by the GA-based library search method. The Monte Carlo simulation result shows that the proposed mixed library-based search method can efficiently find the most adequate set f_1^* containing three different promoter-RBS components and a control inducer concentration to achieve the desired design specifications (see Figure 7.4). Clearly, the fluorescence level of the biological low-pass filter can robustly and optimally match the desired low-pass I/O response despite the intrinsic parameter fluctuations and environmental disturbances.

Similarly, to design a biological filter with a high-pass I/O response as shown in Figure 7.2C, another example of a biological filter circuit based on the gene circuit topology is provided in Figure 7.5. Each module has both

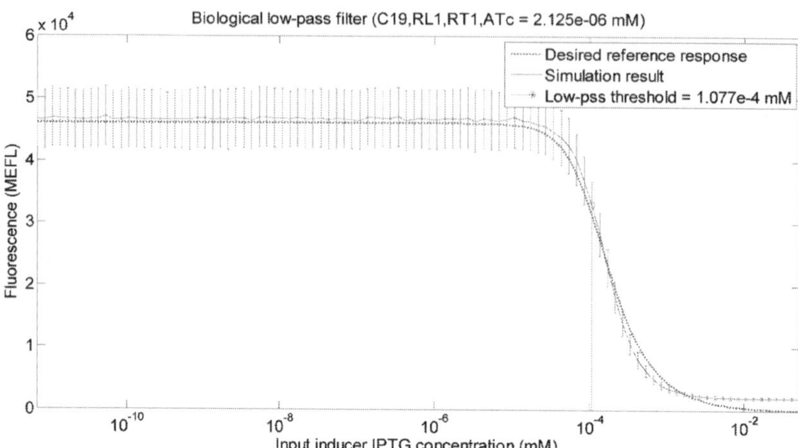

Figure 7.4. Monte Carlo simulation result of a biological low-pass filter displayed as the average of 1000 runs. By minimizing the cost function in (7.10) for the biological low-pass filter in Figure 7.3, the most adequate set $f_1^* = \{pR_1^*, pR_2^*, pR_3^*, I_2^*\} = \{C_{19}, R_{L1}, R_{T1}, ATc = 2.125 \times 10^{-6} \text{ mM}\}$ is selected from the corresponding libraries. The blue solid line is the Monte Carlo simulation result displayed as the average of 1000 runs. The red dashed line is the desired I/O response generated by equation (7.8). The red solid line indicates the low-pass threshold of the biological filter.

Color image of this figure appears in the color plate section at the end of the book.

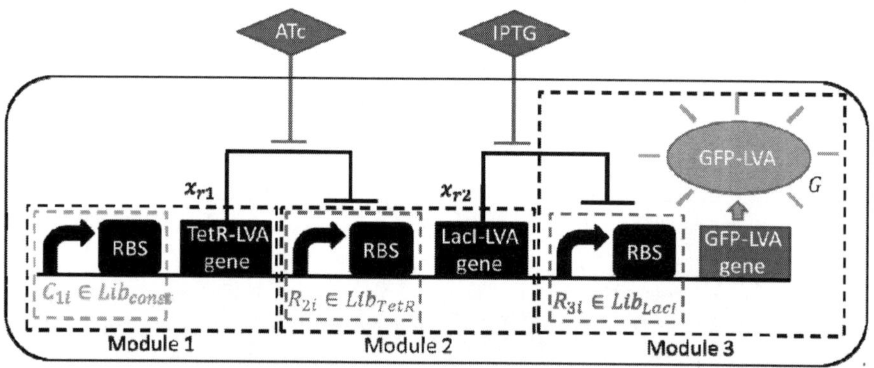

Figure 7.5. A design example of the biological high-pass filter. The biological high-pass filter can be divided into three cascade modules. The first module contains both a constitutive promoter-RBS component C_{1i} and a repressor TetR-LVA gene. The second module contains a TetR-regulated promoter-RBS component R_{2i} and a repressor LacI-LVA gene, and the third module contains a LacI-regulated promoter-RBS component R_{3i} and a reporter protein GFP-LVA gene.

Color image of this figure appears in the color plate section at the end of the book.

a promoter-RBS component selected from the corresponding promoter-RBS libraries and a specified gene downstream of the promoter. The first module contains a constitutive promoter-RBS component C_{1i} selected from the promoter-RBS library Lib_{const} and a TetR-LVA gene. The second module contains a TetR-regulated promoter-RBS component R_{2i} selected from the promoter-RBS library Lib_{TetR} and a LacI-LVA gene. The third module contains a LacI-regulated promoter-RBS component R_{3i} selected from the promoter-RBS library Lib_{LacI} and a GFP-LVA gene. The external inducers, ATc and IPTG, are used to regulate repressor activities. The steady-state model for the biological filter in Figure 7.5 is similar to (7.3). It has a set of promoter-RBS components C_{1i}, R_{2i}, and R_{3i} selected from the promoter-RBS libraries Lib_{const}, Lib_{TetR}, and Lib_{LacI}, respectively. In particular,

$$\{S_{1u,i}\} \in Lib_{const}, \{S_{2u,i}, S_{2l,i}\} \in Lib_{TetR}, \{S_{3u,i}, S_{3l,i}\} \in Lib_{LacI} \tag{7.11}$$

In this design case, x_{r1ss} and x_{r2ss} denote steady-state concentrations of TetR-LVA and LacI-LVA, respectively. I_1 and I_2 denote the external inducer concentrations of ATc and IPTG, respectively. The maximum and minimum strengths $\{S_{2u,i}, S_{2l,i}\}$ of the second promoter-RBS component R_{2i} are selected from the promoter-RBS library Lib_{TetR}. The maximum and minimum strengths $\{S_{3u,i}, S_{3l,i}\}$ of the third promoter-RBS component R_{3i} are selected from the promoter-RBS library Lib_{LacI}. According to the gene circuit topology in Figure 7.5 and the steady-state model in (7.3), if the inducer IPTG is taken as an input to the filter and the inducer ATc is taken as an external control, the high-pass I/O response is obtained. Next, the design specifications for a biological high-pass filter, with a desired high-pass I/O response are prescribed as follows:

i) A desired high-pass I/O response $G_{r2}(I_2)$ (Figure 7.2C) of a biological high-pass filter with a specified threshold $u_{hc} = 10^{-5}$ mM and maximum fluorescence level $G_{max} = 4.3 \times 10^4$ MEFL, is provided as follows for the biological filter matching (7.5):

$$G_{r2}(I_2) = \frac{4.3 \times 10^4 \times \dfrac{I_2}{10^{-5}}}{\sqrt{1^2 + (\dfrac{I_2}{10^{-5}})^2}} \tag{7.12}$$

ii) The index of Lib_{const} is from 1 to 27, the index of Lib_{TetR} is from 1 to 8 and the index of Lib_{LacI} is from 1 to 8. The concentration of control inducer ATc reneges from 0 mM to 4.337×10^{-6} mM.

iii) The standard deviations of the parameter fluctuations are assumed to be the same as (7.9), and the environmental disturbances ω_i of the transcription and translation processes are independent Gaussian white noises with zero means and unit variances.

iv) The following cost function for the biological high-pass filter needs to be minimized

$$J(f_2^*) = \min_{f_2=\{pR_1,pR_2,pR_3,I_1\}} E \int_0^{I_{2f}} (G_{ss}(f_2,I_2) - G_{r2}(I_2))^2 \, dI_2 \qquad (7.13)$$

By the proposed mixed library-based search method, the most adequate set $f_2^* = \{pR_1^*, pR_2^*, pR_3^*, I_1^*\} = \{C_6, R_{T2}, R_{L5}, ATc = 7.807\times10^{-9} \text{ mM}\}$ for the biological high-pass filter is searched. The Monte Carlo simulation result from the most adequate set f_2^* with 1000 runs can confirm that the proposed mixed library-based search method is feasible in the engineering of the biological high-pass filter (see Figure 7.6).Clearly, the fluorescence level of the biological high-pass filter can robustly and optimally match the desired high-pass I/O response despite the intrinsic parameter fluctuations and environmental disturbances.

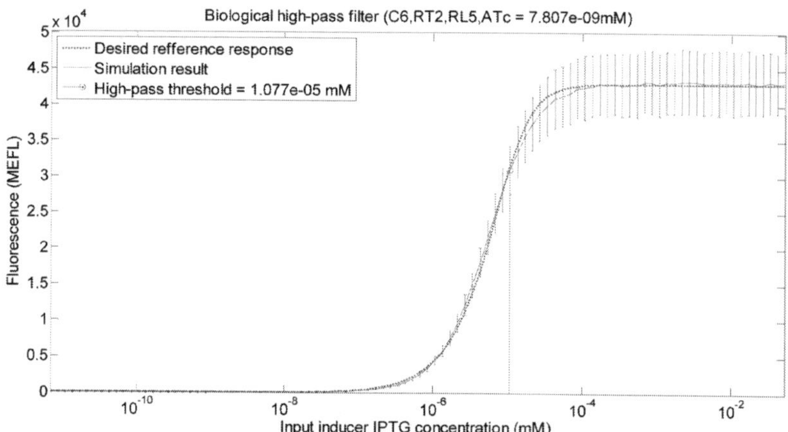

Figure 7.6. Monte Carlo simulation result of a biological high-pass filter displayed as the average of 1000 runs. By minimizing the cost function in (7.13) for the biological high-pass filter in Figure 7.5, the most adequate set $f_2^* = \{pR_1^*, pR_2^*, pR_3^*, I_1^*\} = \{C_6, R_{T2}, R_{L5}, ATc = 7.807\times10^{-9} \text{ mM}\}$ is selected from the corresponding libraries. The blue solid line is the Monte Carlo simulation displayed as the average of 1000 runs. The red dashed line is the desired I/O response generated by equation (7.12). The red solid line indicates the high-pass threshold of the biological filter.

Color image of this figure appears in the color plate section at the end of the book.

7.1.7 External tuning of the biological filter

External tuning of the designed biological filter is possible after the biological filter with a desired I/O response is engineered via the proposed mixed library-based search method, and is discussed as follows. For a biological low-pass filter based on the gene circuit topology in Figure 7.3, the inducers IPTG and ATc are taken as an input and an external control to the filter, respectively. When the most adequate set for the biological low-pass filter is obtained from the above design procedure, we then fix three promoter-RBS components and tune the concentration of external control ATc, thereby demonstrating external tuning of the threshold in our biological filter design. With the most adequate set $f_1^* = \left\{ pR_1^*, pR_2^*, pR_3^*, I_2^* \right\} = \left\{ C_{19}, R_{L1}, R_{T1}, ATc = 2.125 \times 10^{-6} \text{ mM} \right\}$ for the biological low-pass filter, a low-pass I/O response with a specific threshold $u_{lc1} = 1.077 \times 10^{-4}$ mM and a maximum fluorescence expression $G_{max} = 4.702 \times 10^{4}$ MEFL is obtained by simulation. When the concentration of external control ATc is decreased to 4.337×10^{-7} mM, a low-pass I/O response with a specific threshold $u_{lc2} = 3.407 \times 10^{-5}$ mM and a maximum fluorescence expression $G_{max} = 4.660 \times 10^{4}$ MEFL is obtained. When the concentration of external control ATc is increased to 3.470×10^{-6} mM, a low-pass I/O response with a specific threshold $u_{lc3} = 1.708 \times 10^{-4}$ mM and a maximum fluorescence expression $G_{max} = 4.700 \times 10^{4}$ MEFL is obtained. As seen in Figure 7.7, the threshold of the biological low-pass filter is shifted to the right (higher) as the concentration of external control inducer ATc is increased. Likewise, the threshold of the biological low-pass filter is shifted to the left (lower) as the concentration of external control inducer ATc is decreased. The simulation demonstrates the external adjustability of the threshold of the biological low-pass filter with different concentrations of external control, as shown in Figure 7.7. Clearly, the threshold of our biological low-pass filter for IPTG detection can be externally and accurately tuned by changing the concentration of the control inducer ATc.

Similarly, for the biological high-pass filter in Figure 7.5, the inducers IPTG and ATc are taken as an input and an external control to the filter, respectively. If three promoter-RBS components are fixed, then the concentration of external control ATc can be tuned to demonstrate the external adjustability of the threshold in our biological high-pass filter. With the most adequate set $f_2^* = \left\{ pR_1^*, pR_2^*, pR_3^*, I_1^* \right\} = \left\{ C_6, R_{T2}, R_{L5}, ATc = 7.807 \times 10^{-9} \text{ mM} \right\}$ for the biological high-pass filter, a high-pass I/O response with a specific threshold $u_{hc1} = 1.077 \times 10^{-5}$ mM and a maximum fluorescence level $G_{max} = 4.354 \times 10^{4}$ MEFL is obtained by simulation. When the concentration of external control ATc is increased to 8.674×10^{-7} mM, a high-pass I/O response with a specific threshold $u_{hc2} = 1.356 \times 10^{-4}$ mM and a maximum

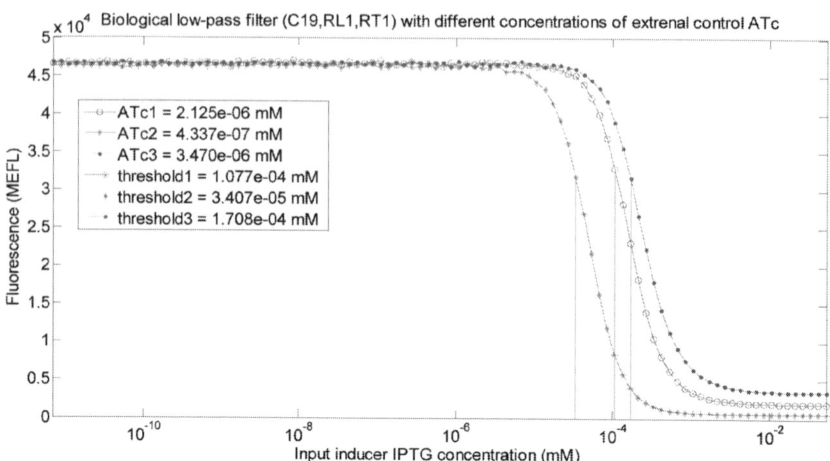

Figure 7.7. Monte Carlo simulations of external adjustability of a biological low-pass filter. The blue line is a low-pass I/O response with a specific threshold $u_{lc1} = 1.077 \times 10^{-4}$ mM and a maximum fluorescence expression $G_{max} = 4.702 \times 10^4$ MEFL based on the most adequate set $f_1^* = \{pR_1^*, pR_2^*, pR_3^*, I_2^*\} = \{C_{19}, R_{L1}, R_{T1}, ATc = 2.125 \times 10^{-6}$ mM$\}$ for the biological low-pass filter. The purple line is a low-pass I/O response with a smaller threshold $u_{lc2} = 3.407 \times 10^{-5}$ mM and a maximum fluorescence expression $G_{max} = 4.660 \times 10^4$ MEFL due to a decrease in the concentration of the external control ATc to 4.337×10^{-7} mM. The green line is a low-pass I/O response with a larger threshold $u_{lc3} = 1.708 \times 10^{-4}$ mM and a maximum fluorescence expression $G_{max} = 4.700 \times 10^4$ MEFL due to an increase in the concentration of the external control ATc to 3.470×10^{-6} mM.

Color image of this figure appears in the color plate section at the end of the book.

fluorescence expression $G_{max} = 4.336 \times 10^4$ MEFL is obtained. When the concentration of external control ATc is further increased to 4.337×10^{-6} mM, a high-pass I/O response with a larger threshold $u_{hc3} = 5.400 \times 10^{-4}$ mM and a maximum fluorescence level $G_{max} = 4.332 \times 10^4$ MEFL is obtained. As seen in Figure 7.8, the threshold of the biological high-pass filter is shifted to the right (higher) as the concentration of control inducer ATc is increased, and vice versa. The simulation demonstrates the external adjustability of the threshold of the biological high-pass filter with different concentrations of external control, as shown in Figure 7.8. Clearly, the threshold of our biological high-pass filter for IPTG detection can be externally and accurately tuned by changing the concentration of the control inducer ATc.

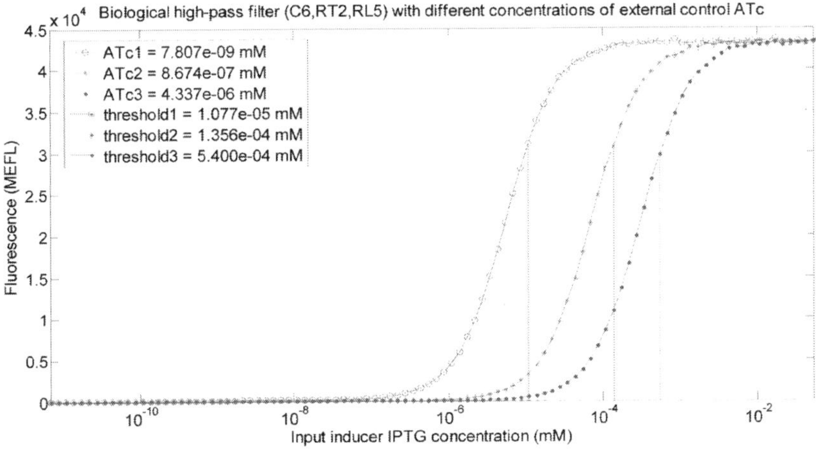

Figure 7.8. Monte Carlo simulations of external adjustability of a biological high-pass filter. The blue line is a high-pass I/O response with a specific threshold $u_{hc1} = 1.077 \times 10^{-5}$ mM and a maximum fluorescence expression $G_{max} = 4.354 \times 10^4$ MEFL based on the most adequate set $f_2^* = \{pR_1^*, pR_2^*, pR_3^*, I_1^*\} = \{C_6, R_{T2}, R_{L5}, ATc = 7.807 \times 10^{-9}$ mM$\}$ for the biological high-pass filter. The purple line is a high-pass I/O response with a larger threshold $u_{hc2} = 1.356 \times 10^{-4}$ mM and a maximum fluorescence expression $G_{max} = 4.336 \times 10^4$ MEFL due to an increase in the concentration of the external control ATc to 8.674×10^{-7} mM. The green line is a high-pass I/O response with a much larger threshold $u_{hc3} = 5.400 \times 10^{-4}$ mM and a maximum fluorescence expression $G_{max} = 4.332 \times 10^4$ MEFL obtained by further increasing the concentration of external control ATc to 4.337×10^{-6} mM.

Color image of this figure appears in the color plate section at the end of the book.

7.2 Systematic Design Methodology for Robust Genetic Transistor

In this section, a robust genetic transistor design example is given with prescribed I/O characteristics of amplification or switching through the external inducer. Before the construction of the synthetic genetic transistor, we introduce the simply operation of an electronic transistor.

7.2.1 The operation of an electronic transistor

A transistor is a semiconductor device used for different functions in electronic circuits, such as signal amplification and switching. It has the advantages of low volume, high efficiency, a long life-span and high speed switching. Two major types of transistor are the bipolar junction transistor

(BJT) and metal-oxide-semiconductor field-effect transistor (MOSFET). The functions of these two types of transistors are similar. The electronic signal can be amplified linearly or switched by controlling the bias voltage. Considering the BJT for example, a three-terminal npn BJT has three terminal connections which are called the emitter, base and collector, as shown in Figure 7.9A (Neamen 2007). The complete npn BJT circuit in the common-emitter configuration is shown in Figure 7.9B, in which the electronic signal from base terminal i_B can be amplified linearly or switched through BJT by controlling the voltage between the collector and emitter terminals V_{CE} in the linear region larger than 0.3V or saturation region smaller than 0.3V. When V_{CE} is controlled in the linear region, i_B is amplified linearly. The signal output from collector terminal i_C is almost equal to $\beta \cdot i_B$ and the variation in input voltage V_{in} is also amplified reversely and linearly, where the parameter β is the common-emitter current gain within the range of 50 to 300. On the other hand, when V_{CE} is controlled in the saturation region, the output signal will be switched such that a high level signal will be switched into a low level signal and vice versa (Neamen 2007, Varadarajan and Del Vecchio 2009). The mathematical models to describe the behaviors of an npn BJT circuit in the common-emitter configuration are shown as follows.

In the cut-off region: $V_{in} \leq V_{BE(on)}$

$$V_o = V_{CC}$$

(7.14)

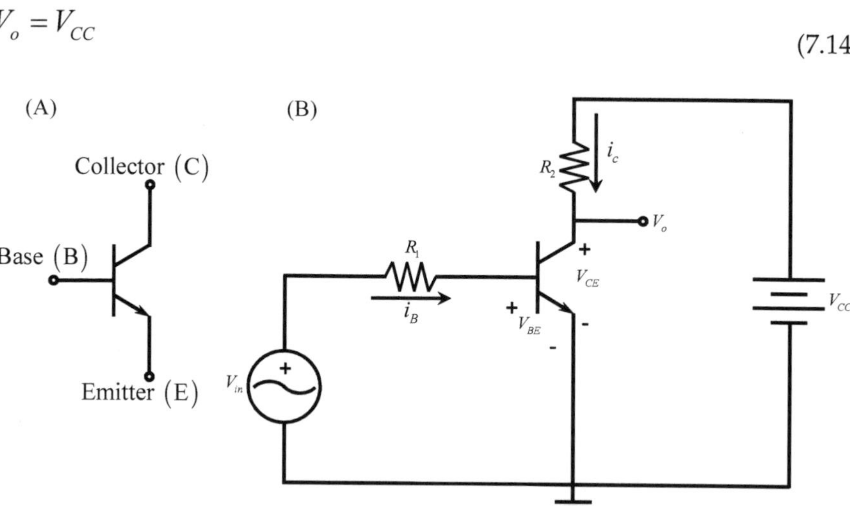

Figure 7.9. The npn bipolar junction transistor. (A) The npn BJT circuit symbol. BJT contains three terminal connections, i.e., the so-called Base, Collector and Emitter. (B) The npn BJT circuit in the common-emitter configuration.

In the forward-active region (linear region): $V_{in} > V_{BE(on)}$

$$V_o = -\beta \frac{R_2}{R_1}\left(V_{in} - V_{BE(on)}\right) + V_{CC} \tag{7.15}$$

In the saturation region: $V_{in} \geq \dfrac{R_1}{\beta \cdot R_2}\left(V_{CC} - V_{CE(sat)}\right) + V_{BE(on)}$

$$V_o = V_{CE(sat)} \tag{7.16}$$

where $V_{BE(on)}$ and V_{CC} denote the turn-on and bias voltages of the npn BJT circuit, respectively, and $V_{CE(sat)}$ denotes the voltage between collector and emitter in saturation. The current-voltage and voltage I/O characteristics of the npn BJT circuit in the common-emitter configuration are shown in Figure 7.10A and B, respectively.

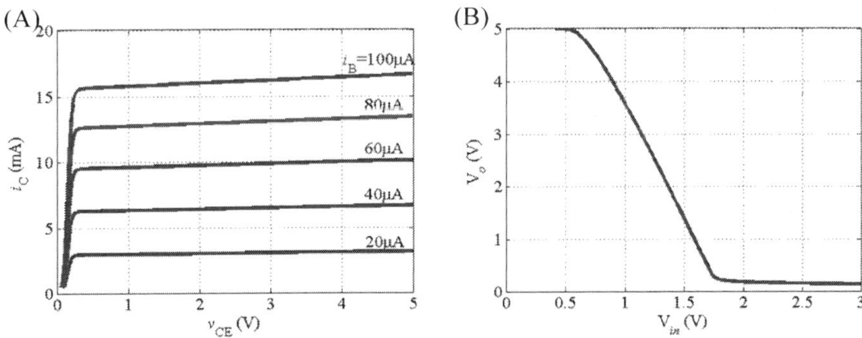

Figure 7.10. The current-voltage characteristic and voltage I/O characteristic of BJT in the common-emitter configuration. The BJT circuit is shown in Figure 7.9B, and its characteristics are simulated by the PSpice with a standard 2N3904 transistor from PSpice library. (A) The current-voltage characteristic: The base current i_B changes from $20\mu A$ to $100\mu A$. (B) The voltage I/O characteristic: Set R_1=5kΩ, R_2=150kΩ, V_{CC}=5V, $V_{BE(on)}$=0.7V, $V_{CE(sat)}$=0.2V and the input voltage V_{in} from 0 V to 3 V. For $V_{in} \leq 0.7$V, the transistor is cut off as shown in (7.14); for 0.7 V<V_{in}<1.9V , the transistor is in the region of the linear amplification as shown in (7.15); and for $V_{in} \geq 1.9$V, the transistor is in the saturation region as shown in (7.16).

7.2.2 Construction of the genetic transistor

A synthetic genetic transistor is shown in Figure 7.11A. The transistor is constructed to obtain the output protein concentration $x_{protein}$ of the transistor for amplification or switching behavior. The genetic transistor consists of the repressor-regulated promoter-RBS component c_3 and a repressor coding gene. The input repressor $x_{repressor2}$ to the genetic transistor is controlled by

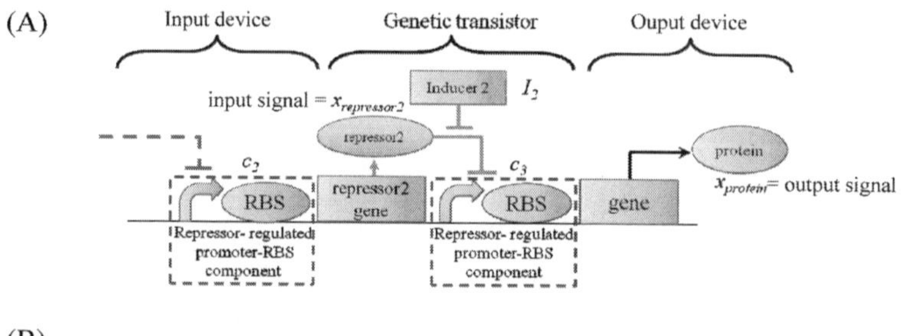

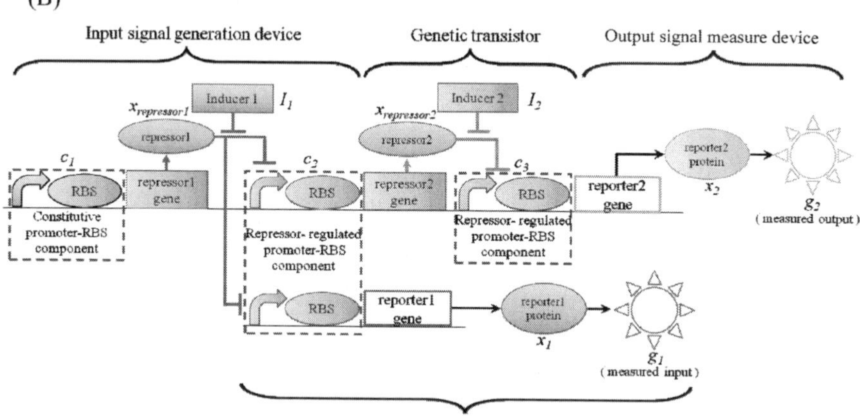

Figure 7.11. The representation of synthetic genetic transistor circuit with different measurement devices. (A) The synthetic genetic transistor consists of a repressor regulated promoter-RBS component c_3 and repressor2 gene $x_{repressor2}$ is the input signal and $x_{protein}$ is the output signal in the genetic transistor. (B) For measuring the I/O characteristics of synthetic genetic transistor circuit, the reporter1 gene and reporter2 gene are embedded at the downstream of repressor-regulated promoter-RBS component c_2 and the additional promoter-RBS component c_3 respectively. Then input protein concentration x_1 can be measured by input fluoresce by g_1 and output fluorescence g_2 of output protein concentration x_2 can be measured by output fluoresce g_2, respectively. In addition, for the convenience of controlling the input, the repressor1 gene is constructed with the constitutive promoter-RBS component c_1 to control the kinetic strength of c_2 and the corresponding inducer I_1 is used to control the fluorescence g_1 of input protein concentration x_1. It is found at steady state that input/output = $x_{repressor2}/x_{protein} \approx x_1/x_2 \approx g_1/g_2$. Therefore the I/O characteristic of $x_{repressor2}/x_{protein}$ in genetic transistor at steady state can be represented by the g_1/g_2 ratio.

Color image of this figure appears in the color plate section at the end of the book.

the repressor-regulated promoter-RBS component c_2, which is regulated by the corresponding repressor. The input repressor $x_{repressor2}$ will form the complex and restrict the production of output protein $x_{protein}$ by binding the corresponding repressor-regulated promoter-RBS component c_3 to decrease its kinetic strength. However, when the inducer is added, this inducer will bind input repressor $x_{repressor2}$ and prevent it from binding to the repressor-regulated promoter-RBS component c_3. Then, both the kinetic strength of repressor-regulated promoter-RBS component c_3 and the production of output protein $x_{protein}$ will increase. The dynamic model of a genetic transistor is described as follows:

$$\dot{x}_{repressor2}(t) = P_{repressor}\left(S_{u,c_2}, S_{l,c_2}, 0, 0\right) - \left(\mu + \gamma_{repressor2}\right)x_{repressor2}(t)$$

$$\dot{x}_{protein}(t) = P_{repressor}\left(S_{u,c_3}, S_{l,c_3}, x_{repressor2}, I_2\right) - \left(\mu + \gamma_{protein}\right)x_{protein}(t) \tag{7.17}$$

where $x_{repressor2}$ and $x_{protein}$ denote the concentrations of input repressor2 and the output protein of the genetic transistor, respectively, and $\gamma_{protein}$ denotes the degradation rate of the protein.

However, the protein concentration is difficult to directly measure and quantify. To determine characteristics of the synthetic genetic transistor, a genetic transistor with measurement circuit is constructed as shown in Figure 7.11B. In Figure 7.11B, we construct an additional repressor-regulated promoter-RBS component c_2 so that the input reporter protein x_1 can be measured by input fluorescence g_1 and the output reporter protein x_2 can be measured by output fluorescence g_2. Additionally, for the convenience of the input regulation, we construct an input signal generation device with the concentration of inducer I_1 to control the input g_1 of the genetic transistor circuit. Then, the dynamic model of a synthetic genetic transistor circuit with I/O measurement devices under environmental disturbances is described by the following set of equations (Lee et al. 2013):

$$\begin{cases} \dot{x}_{repressor1}(t) = P_{const}\left(S_{u,c_1}\right) - \left(\mu + \gamma_{repressor1}\right)x_{repressor1}(t) + v_1(t) \\[4pt] \dot{x}_{repressor2}(t) = P_{repressor}\left(S_{u,c_2}, S_{l,c_2}, x_{repressor1}, I_1\right) - \left(\mu + \gamma_{repressor2}\right)x_{repressor2}(t) + v_2(t) \\[4pt] \dot{x}_1(t) = P_{repressor}\left(S_{u,c_2}, S_{l,c_2}, x_{repressor1}, I_1\right) - \left(m_1 + \mu + \gamma_{im,x_1}\right)x_1(t) + v_3(t) \\[4pt] \dot{g}_1(t) = m_1 \cdot x_1(t) - \left(\mu + \gamma_{m,x_1}\right)g_1(t) + v_4(t) \\[4pt] \dot{x}_2(t) = P_{repressor}\left(S_{u,c_3}, S_{l,c_3}, x_{repressor2}, I_2\right) - \left(m_2 + \mu + \gamma_{im,x_2}\right)x_2(t) + v_5(t) \\[4pt] \dot{g}_2(t) = m_2 \cdot x_2(t) - \left(\mu + \gamma_{m,x_2}\right)g_2(t) + v_6(t), \end{cases} \tag{7.18}$$

$$c_1 \in Lib_{const},\ c_2 \text{ and } c_3 \in Lib_{repressor}$$

where m_1 and m_2 denote the maturation rates of reporter1 x_1 and reporter2 x_2, respectively, and $v_i(t)$, $i = 1, 2, \cdots, 6$ denote the noises.

To explore the I/O characteristics of a synthetic genetic transistor with the function of amplification or switching, the steady state model of (7.18) is given by (Lee et al. 2013)

$$\begin{cases}
x_{repressor1} = P_{const}\left(S_{u,c_1}\right)\Big/\left(\mu + \gamma_{repressor1}\right) + v_{s_1} \\[6pt]
x_{repressor2}(I_1) = P_{repressor}\left(S_{u,c_2}, S_{l,c_2}, x_{repressor1}, I_1\right)\Big/\left(\mu + \gamma_{repressor2}\right) + v_{s_2} \\[6pt]
x_1(I_1) = P_{repressor}\left(S_{u,c_2}, S_{l,c_2}, x_{repressor1}, I_1\right)\Big/\left(m_1 + \mu + \gamma_{im,x_1}\right) + v_{s_3} \\[6pt]
g_1(I_1) = m_1 \cdot x_1 \Big/ \left(\mu + \gamma_{m,x_1}\right) + v_{s_4} \\[6pt]
x_2(I_1, I_2) = P_{repressor}\left(S_{u,c_3}, S_{l,c_3}, x_{repressor2}, I_2\right)\Big/\left(m_2 + \mu + \gamma_{im,x_2}\right) + v_{s_5} \\[6pt]
g_2(I_1, I_2) = m_2 \cdot x_2 \Big/ \left(\mu + \gamma_{m,x_2}\right) + v_{s_6},
\end{cases} \tag{7.19}$$

$$c_1 \in Lib_{const}, c_2 \text{ and } c_3 \in Lib_{repressor}$$

where v_{s_i}, $i = 1, 2, \cdots, 6$ denote the noises at the steady state.

From (7.19), if $m_1 \approx m_2$, $\gamma_{m,x_1} \approx \gamma_{m,x_2}$ and $\gamma_{im,x_1} \approx \gamma_{im,x_2}$, then the I/O characteristic can be regarded as input/output $= x_{repressor2}/x_{protein} \approx x_1/x_2 \approx g_1/g_2$, i.e., we could use the x_1/x_2 or g_1/g_2 ratio to replace the I/O characteristic of the synthetic genetic transistor. Further, the I/O characteristic can be controlled and regulated by the selection of promoter-RBS components c_3 and inducer concentration I_2. Therefore, we need to define the I/O characteristic of synthetic genetic transistor circuits to design a genetic transistor with the desired I/O characteristic. This is done as follows:

$$y_{ss}(c_3, I_2, g_1) = g_2(c_3, I_2, g_1), \quad g_1 \in [g_{1e}, g_{1n}] \tag{7.20}$$

where $y_{ss}(c_3, I_2, g_1)$ denotes the I/O response of the synthetic genetic transistor circuit between input signal g_1 and output signal g_2, and g_{1e} and g_{1n} denote the lower bound and upper bound of g_1.

In Figure 7.11B, promoter-RBS components c_1 and c_2 can be selected to control input signals $x_{repressor1}$ and $x_{repressor2}(I_1)$ in (7.19). In general, genetic components are inherently uncertain in the biological system as a result of gene expression noises in transcription or translation processes, thermal fluctuations, DNA mutations, evolutions and parameter estimation errors (Chen and Wu 2009, Chen et al. 2011). Hence, we model the uncertain kinetic strengths of promoter-RBS components, degradation rate of proteins and transcription/translation rates as stochastic processes in the following model:

$$\begin{aligned}
&S_{u,c_1} \to S_{u,c_1} + \Delta S_{u,c_1} n_1(t),\; S_{u,c_2} \to S_{u,c_2} + \Delta S_{u,c_2} n_2(t),\; S_{l,c_2} \to S_{l,c_2} + \Delta S_{l,c_2} n_2(t), \\[4pt]
&S_{u,c_3} \to S_{u,c_3} + \Delta S_{u,c_3} n_3(t),\; S_{l,c_3} \to S_{l,c_3} + \Delta S_{l,c_3} n_3(t), \\[4pt]
&\gamma_{repressor1} \to \gamma_{repressor1} + \Delta\gamma_{repressor1} n_1(t),\; \gamma_{repressor2} \to \gamma_{repressor2} + \Delta\gamma_{repressor2} n_2(t), \\[4pt]
&\gamma_{im,x_1} \to \gamma_{im,x_1} + \Delta\gamma_{im,x_1} n_2(t),\; \gamma_{m,x_1} \to \gamma_{m,x_1} + \Delta\gamma_{m,x_1} n_2(t), \\[4pt]
&\gamma_{im,x_2} \to \gamma_{im,x_2} + \Delta\gamma_{im,x_2} n_3(t),\; \gamma_{m,x_2} \to \gamma_{m,x_2} + \Delta\gamma_{m,x_2} n_3(t),\; m_1 \to m_1 + \Delta m_1 n_2(t), \\[4pt]
&m_2 \to m_2 + \Delta m_2 n_3(t),\; \mu \to \mu + \Delta\mu n_1(t)
\end{aligned} \tag{7.21}$$

where $\Delta S_{u,c_1}$, $\Delta S_{u,c_2}$, $\Delta S_{l,c_2}$, $\Delta S_{u,c_3}$, $\Delta S_{l,c_3}$, $\Delta \gamma_{repressor1}$, $\Delta \gamma_{repressor2}$, $\Delta \gamma_{im,x_1}$, $\Delta \gamma_{m,x_1}$, $\Delta \gamma_{im,x_2}$, $\Delta \gamma_{m,x_2}$, Δm_1, Δm_2 and $\Delta \mu$ denote the standard deviations of stochastic parameters and $n_i(t)$, $i = 1, 2, 3$ denote Gaussian noises with zero mean and unit variance. Therefore, $\Delta S_{u,c_1}$, $\Delta S_{u,c_2}$, $\Delta S_{l,c_2}$, $\Delta S_{u,c_3}$, $\Delta S_{l,c_3}$, $\Delta \gamma_{repressor1}$, $\Delta \gamma_{repressor2}$, $\Delta \gamma_{im,x_1}$, $\Delta \gamma_{m,x_1}$, $\Delta \gamma_{im,x_2}$, $\Delta \gamma_{m,x_2}$, Δm_1, Δm_2 and $\Delta \mu$ denote the deterministic parts of parameter variations and $n_i(t)$, $i = 1, 2, 3$ denote different random fluctuation sources. For robust design of the genetic transistor circuit, these parameter fluctuations in (7.21) will hence be considered in the design procedure so that the synthetic genetic transistor can tolerate these kinds of parameter fluctuations *in vivo*.

With fixed concentration of inducer I_2, we expect that the input signal g_1/output signal g_2 (I/O) characteristics of the synthetic genetic transistor in (7.20) would be similar to the voltage I/O characteristics of the electronic transistor shown in Figure 7.10B. When the inducer concentration I_1 increases, the kinetic strength of promoter-RBS component c_2 increases along with the fluorescence of the input signal g_1, which means that the repressor concentration $x_{repressor2}$ increases. Due to the fixed concentration of inducer I_2, the redundant repressors $x_{repressor2}$, which are not bound by the inducer I_2, will repress the promoter-RBS component c_3, and the fluorescence of output signal g_2 will decrease. Therefore, the I/O characteristic of the synthetic genetic transistor is similar to Figure 7.10B. Additionally, from Figure 7.10B, we see that if input signal is in the operation range of linear amplification, the input signal would be inversely amplified.

Now, consider the alternative viewpoint, i.e., the voltage I/O characteristics of an electronic transistor. When R_2/R_1 increases, the reverse amplification gain will become large and the operation region of linear amplification will narrow as shown in (7.14)–(7.16) and Figure 7.12. In the synthetic genetic transistor, we expect that when the concentration of inducer I_2 changes as per the R_2/R_1 ratio in (7.15)–(7.16), the I/O characteristics would be similar to the voltage I/O characteristics of electronic transistor in Figure 7.12. Due to different concentrations of inducer I_2, the effect of the inducer on the input repressor can vary. When the inducer concentration I_2 decreases, the I/O characteristics would sharpen, so the reverse amplification gain becomes large in the operation region of linear amplification.

Finally, when R_2/R_1 is large enough in (7.15)–(7.16), the operation region of linear amplification will become too narrow and result in a sharp change in this region. Correspondingly with a synthetic genetic transistor, when the inducer concentration I_2 is low enough, the input signal g_1 will produce a small variation, and the output signal g_2 will have an acute change like a switch. Therefore, according to the analysis above, we could obtain varying reverse amplification gains and switch levels by changing the concentration of inducer I_2.

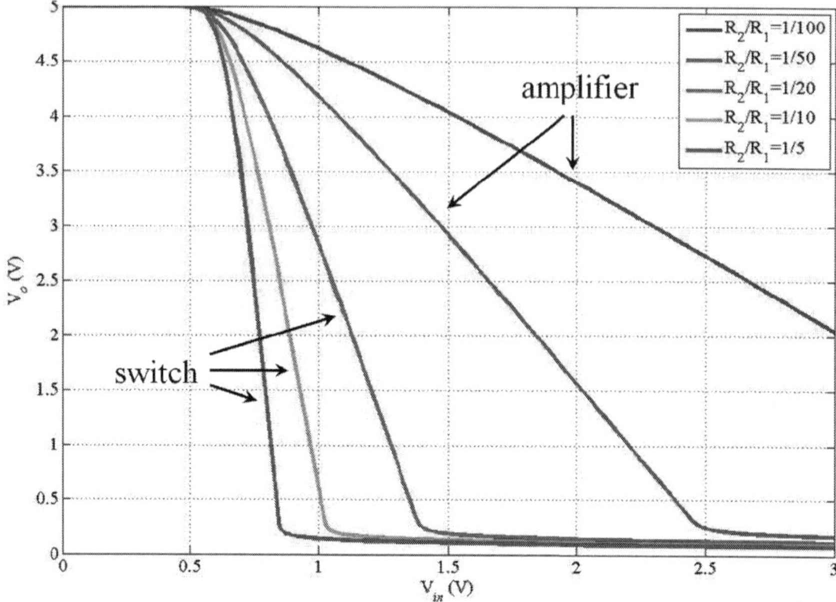

Figure 7.12. The voltage I/O characteristics of the common-emitter circuit for different R_2/R_1 ratio. The circuit is shown in Figure 7.9B, and simulated by the PSpice with a standard 2N3904 transistor from PSpice library. The voltage I/O characteristics are simulated by changing the R_2/R_1 ratio. According to (7.15), when the R_2/R_1 ratio becomes large, the voltage I/O characteristic is much sharper as the amplifier in linear region. And when the R_2/R_1 ratio is large enough, the voltage I/O characteristic will be like a switch.

Color image of this figure appears in the color plate section at the end of the book.

7.2.3 Systematic design of a genetic transistor based on design specification

According to the above analysis in Figure 7.11B, we can obtain different reverse amplification gains or switch behaviors via regulation of different concentrations of inducer I_2. Additionally, due to the output signal g_2 being under the controlled by promoter-RBS component c_3, we could change the output range by selecting different repressor-regulated promoter-RBS components c_3 from the repressor-regulated promoter-RBS libraries. In this way, we can control the I/O characteristics of a synthetic genetic transistor to obtain different reverse amplification gains or switch levels by choosing different concentrations of inducer I_2 and selecting different repressor-regulated promoter-RBS components c_3 from the repressor-regulated promoter-RBS libraries.

In Figure 7.11B, the input signal generation device consists of a constitutive promoter-RBS component c_1, and a repressor-regulated promoter-RBS component c_2 and an inducer I_1. The constitutive promoter-RBS component c_1 is selected to produce the input repressor continually. Further, for convenience of design, the repressor-regulated promoter-RBS component c_2 is selected from the corresponding promoter-RBS library to have sufficient kinetic strength to obtain an adequate maximum regulation range of input signal regulated by inducer I_1. However, the operation region of linear amplification is still limited in the amplifier design of genetic transistor, and the input signal range might not be fully contained in the operation region of linear amplification. Therefore, the input signal range should be considered in relation to the design purpose. In the procedure of amplifier design, the input operation range $g_1 \in [g_{1,l}, g_{1,u}]$ can be set by transforming the inducer concentration I_1 into the input fluorescence according to (7.18) or (7.19) as follows

$$\text{Input operation range}: I_1 \in [I_{1,l}, I_{1,u}] \Rightarrow g_1 \in [g_{1,l}, g_{1,u}] \tag{7.22}$$

where I_1 and g_1 denote the inducer concentration and input fluorescence, respectively, $I_{1,l}$ and $I_{1,u}$ denote the lower and upper bound of inducer concentrations, respectively, and $g_{1,l}$ and $g_{1,u}$ denote the lower and upper bound of input fluorescence respectively.

Note that, in the future, when the promoter-RBS libraries are large enough, the promoter-RBS components c_1 and c_2 can be designed and selected to match the input operation range. However, due to the limited size of our promoter-RBS libraries and for the convenience of design, we will select the repressor-regulated promoter-RBS component c_2 from the corresponding promoter-RBS library.

From the above analysis, the design purpose of an amplifier will lead to the selection of a suitable repressor-regulated promoter-RBS component c_3 from the repressor-regulated promoter-RBS libraries and concentration of inducer I_2, so that the I/O characteristics of the synthetic genetic transistor in (7.20) in a specific input range $g_1 \in [g_{1,l}, g_{1,u}]$ can match the desired I/O response similar to (7.15), i.e.,

$$y_d(g_1) = gain \cdot (g_1 - g_{1,l}) + g_{2,u}, \quad g_1 \in [g_{1l}, g_{1u}] \tag{7.23}$$

where $g_{1,l}$ and $g_{2,u}$ denote the lower bound of input fluorescence g_1 and upper bound of output fluorescence g_2, respectively, and *gain* denotes the amplification gain of the genetic transistor.

On the other hand, the switching behavior will occur when the input signal has a small variation (see Figure 7.12), i.e., a high level signal can be switched into a low level signal and vice versa. In the switching behavior of synthetic genetic transistor, each promoter-RBS component has its own basal level. Thus, when the input signal increases, the output signal will rapidly decrease to the basal level. Therefore, the desired I/O response of a switch is described as follows (Lee et al. 2013):

$$y_d(g_1) = L_s + \frac{(H_s - L_s)}{1 + (g_1/g_t)^2}, \quad g_1 \in [g_{1l}, \ g_{1u}] \tag{7.24}$$

where H_s and L_s denote the high level and low level of switching, respectively, and g_t denotes the transition point of input fluorescence. Moreover, the input signal range of I/O characteristics of the switch can be set by (7.22).

Finally, for matching the desired I/O response of an amplifier or switch, the genetic algorithm (GA) is employed to select an adequate repressor-regulated promoter-RBS component c_3 in the repressor-regulated promoter-RBS libraries and the concentration of inducer I_2 to minimize the following cost function (Wu et al. 2011, Lee et al. 2013):

$$J(c_3, I_2) = \min_{c_3 \in Lib_{repressor}, \ I_2 \in [I_{2,l}, I_{2,u}]} \mathrm{E} \int_{g_{1,l}}^{g_{1,u}} (y_{ss}(c_3, I_2, g_1) - y_d(g_1))^2 dg_1 \tag{7.25}$$

To summarize the above design procedure of a biological amplifier and switch, a genetic transistor design procedure of by the promoter-RBS library searching method using GA is proposed as follows (Wu et al. 2011, Lee et al. 2013):

1. Construct the genetic transistor circuit such as in Figure 7.11A.
2. Build the mathematical model in (7.18) or (7.19).
3. Provide the design specification of amplifier with the desired I/O response as in (7.22) and (7.23) or switch with the desired I/O response as in (7.22) and (7.24).
4. Provide the standard deviations of parameter fluctuations and environmental disturbances to be tolerated *in vivo* in (7.21).
5. Minimize the cost function $J(c_3, I_2)$ in (7.25) by selecting an optimal set $\{c_3, I_2\}$ via GA.

Based on the design procedure of a genetic transistor using the promoter-RBS library searching method with GA, the promoter-RBS component c_3 is selected from the corresponding repressor-regulated promoter-RBS library and the inducer concentration I_2 is selected within $[I_{2,l}, I_{2,u}]$, while the cost function is calculated in each iteration of the selection process. Then, GA would select the most adequate promoter-RBS component c_3 from the corresponding repressor-regulated promoter-RBS library and inducer concentration $I_2 \in [I_{2,l}, I_{2,u}]$ to minimize the cost function.

Notice that in the *in vivo* experiment, due to different reporters with different degradation rates, the I/O characteristic will be affected by choosing different reporters to measure the input and output signal. To simplify the design problem, we will choose the same reporter to measure the input signal g_1 and output signal g_2. Therefore, we only present the effect of changing promoter-RBS component c_3 and concentration of inducer I_2 in Figure 7.11B. However, at present, the instrument cannot measure the fluorescence of the input and output signals with the same reporter protein. Therefore, in order to measure the fluorescence of input and output signals with the same reporter protein *in vivo*, the synthetic genetic transistor in Figure 7.11B is divided into two parts with the same reporter protein, as shown in Figure 7.13A and B. The portion shown in Figure 7.13A can be used to measure the concentration of input signal, while that shown in Figure 7.13B can be used to measure the concentration of output signal. The two separate genetic circuits in Figure 7.13A and B will be constructed in two cells, which will grow under the same conditions and can be measured simultaneously.

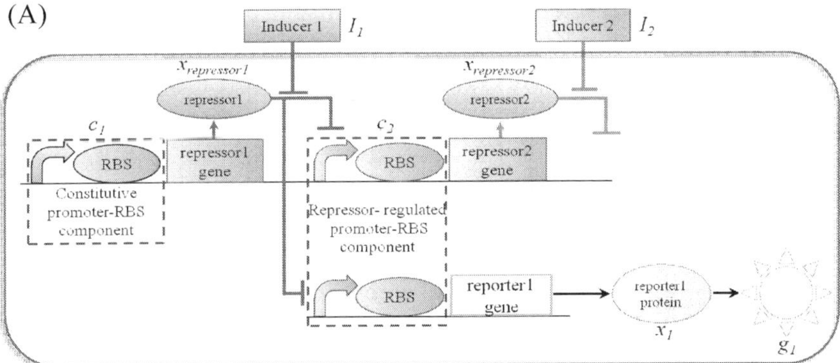

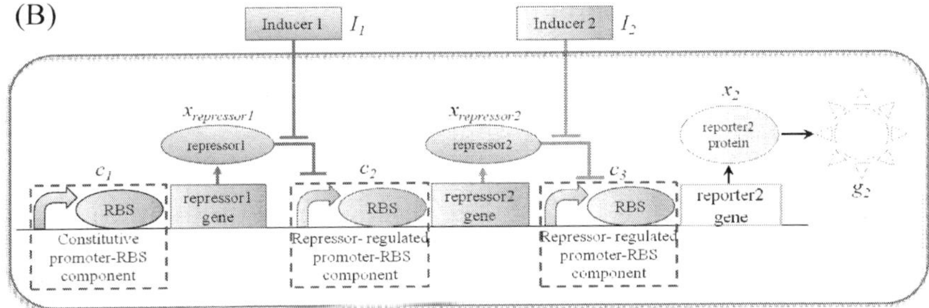

Figure 7.13. Input and output signal measurement of synthetic genetic transistor *in vivo*. For measuring the input fluorescence g_1 and output fluorescence g_2 with the same reporter gene, the synthetic genetic transistor circuit is divided into two parts in (A) and (B). These two parts will be constructed into different cells, respectively. Then, the cells will growth in the same condition and are measured at the same time. (A) The measurement of the input fluorescence g_1 of synthetic genetic transistor. (B) The measurement of the output fluorescence g_2 of synthetic genetic transistor.

Color image of this figure appears in the color plate section at the end of the book.

7.2.4 Amplifier design example of synthetic genetic transistor

Consider the synthetic genetic transistor in Figure 7.14A. For the convenience of *in vivo* measurement of I/O characteristics of a synthetic genetic transistor circuit, the genetic transistor input/output signal measurement devices can be reconstructed as shown in Figure 7.14B. The genetic transistor in Figure 7.14B has three promoter-RBS components, c_1, c_2 and c_3, to be selected from promoter-RBS libraries, Lib_{const}, Lib_{LacI} and Lib_{TetR}, respectively. The input signal generation device is through the inducer IPTG which binds to the LacI protein to prevent it from binding the promoter-RBS component c_2, thus controlling of the fluorescence g_1 of input signal x_1. In the main signal

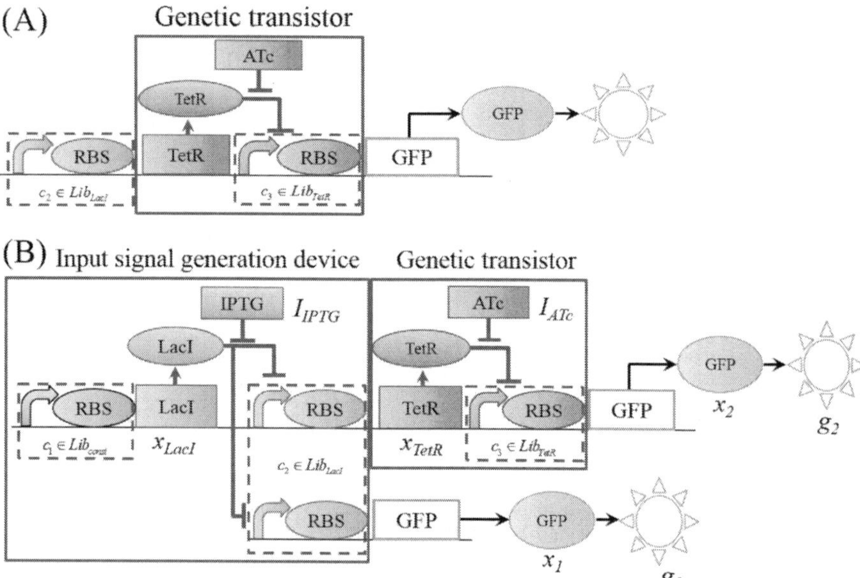

Figure 7.14. The genetic transistor design example based on promoter-RBS libraries. (A) The genetic transistor consists of repressor-regulated promoter-RBS component, $c_3 \in Lib_{Tet}$, TetR protein coding gene and inducer, ATc. (B) The complete genetic transistor with measurement circuit consists of three promoter-RBS components, c_1, c_2 and c_3, selected from promoter-RBS libraries, Lib_{const}, Lib_{LacI} and Lib_{TetR}, respectively. LacI protein x_{LacI} represses the strength of promoter-RBS component c_2 to decrease the concentration of TetR protein x_{TetR} and the fluorescence g_1 of GFP x_1. The inducer I_{IPTG} is added to bind LacI protein x_{LacI} to restrain LacI protein x_{LacI} from repressing the strength of promoter-RBS component c_2. And TetR protein x_{TetR} represses the strength of promoter-RBS component c_3 to decrease the fluorescence g_2 of GFP x_2. The inducer I_{ATc} is added to bind TetR protein x_{TetR} to restrain TetR protein x_{TetR} from repressing the strength of promoter-RBS component c_3.

Color image of this figure appears in the color plate section at the end of the book.

processor of the genetic transistor, the inducer ATc is added to bind to the TetR protein, controlling the effect of the TetR protein in a bid to regulate the promoter-RBS component c_3. Then, the dynamic model in Figure 7.14B is described as follows (Lee et al. 2013):

$$
\begin{cases}
\dot{x}_{LacI}(t) = P_{const}\left(S_{u,c_1}\right) - \left(\mu + \gamma_{LacI}\right)x_{LacI}(t) \\[4pt]
\dot{x}_{TetR}(t) = P_{LacI}\left(S_{u,c_2}, S_{l,c_2}, x_{LacI}, I_{IPTG}\right) - \left(\mu + \gamma_{TetR}\right)x_{TetR}(t) \\[4pt]
\dot{x}_1(t) = P_{LacI}\left(S_{u,c_2}, S_{l,c_2}, x_{LacI}, I_{IPTG}\right) - \left(m_1 + \mu + \gamma_{im,x_1}\right)x_1(t) \\[4pt]
\dot{g}_1(t) = m_1 \cdot x_1(t) - \left(\mu + \gamma_{m,x_1}\right)g_1(t) \\[4pt]
\dot{x}_2(t) = P_{TetR}\left(S_{u,c_3}, S_{l,c_3}, x_{TetR}, I_{ATc}\right) - \left(m_2 + \mu + \gamma_{im,x_2}\right)x_2(t) \\[4pt]
\dot{g}_2(t) = m_2 \cdot x_2(t) - \left(\mu + \gamma_{m,x_2}\right)g_2(t),
\end{cases}
\tag{7.26}
$$

$$c_1 \in Lib_{const},\ c_2 \in Lib_{Lac} \text{ and } c_3 \in Lib_{Tet}$$

where $\gamma_{LacI}, \gamma_{TetR}, \gamma_{im,x_i}$ and $\gamma_{m,x_i}, i = 1,2$, denote the degradation rates of LacI, TetR, immature and mature GFP, respectively while μ and $m_i, i = 1,2$, denote the dilution rate and maturation rates, respectively.

A detailed amplifier design procedure is as follows:

1. First, the promoter-RBS components $\{c_1, c_2\} = \{J_1, L_1\}$ are to be selected to control of the input signal range. The promoter-RBS component c_2 with the greatest kinetic strength is selected from the promoter-RBS library Lib_{LacI} for maximum regulation of the input range.
2. For the dynamic model in (7.26), the steady state model is described as follows (Lee et al. 2013):

$$
\begin{cases}
x_{LacI} = P_{const}\left(S_{u,c_1}\right)\big/\left(\mu + \gamma_{LacI}\right) \\[4pt]
x_{TetR}\left(I_{IPTG}\right) = P_{LacI}\left(S_{u,c_2}, S_{l,c_2}, x_{LacI}, I_{IPTG}\right)\big/\left(\mu + \gamma_{TetR}\right) \\[4pt]
x_1\left(I_{IPTG}\right) = P_{LacI}\left(S_{u,c_2}, S_{l,c_2}, x_{LacI}, I_{IPTG}\right)\big/\left(m_1 + \mu + \gamma_{im,x_1}\right) \\[4pt]
g_1\left(I_{IPTG}\right) = m_1 \cdot x_1(t)\big/\left(\mu + \gamma_{m,x_1}\right) \\[4pt]
x_2\left(I_{IPTG}, I_{ATc}\right) = P_{TetR}\left(S_{u,c_3}, S_{l,c_3}, x_{TetR}, I_{ATc}\right)\big/\left(m_2 + \mu + \gamma_{im,x_2}\right) \\[4pt]
g_2\left(I_{IPTG}, I_{ATc}\right) = m_2 \cdot x_2(t)\big/\left(\mu + \gamma_{m,x_2}\right)
\end{cases}
\tag{7.27}
$$

$$c_1 \in Lib_{const},\ c_2 \in Lib_{Lac} \text{ and } c_3 \in Lib_{Tet}$$

Due to the measurement of the input and output signal using the same GFP, the degradation rates γ_{im,x_1} and γ_{im,x_2} of the immature GFP are both equal to γ_{im}, and similarly, the degradation rates γ_{m,x_1} and γ_{m,x_2} of the mature GFP are both equal to γ_m. The kinetic parameter values for simulation are listed in Table 7.1.

3. The concentration range of inducer I_{IPTG} is from 100 to 200 nM and the input signal range can be obtained by the genetic circuit in Figure 7.14 and according to (7.26) and (7.27) as follows:

Input operation range :

$$I_{IPTG} \in [100,200]\,(\text{nM}) \Rightarrow g_1 \in [5152,12330]\,(\text{MEFL}) \tag{7.28}$$

and we set the amplification gain as -2. Then, the desired I/O response in the given input operation range can be obtained by

$$y_d\left(g_1\right) = -2 \cdot (g_1 - 5152) + 57480, \quad 5152 \le g_1 \le 12330\,(\text{MEFL}) \tag{7.29}$$

4. For robust design *in vivo*, suppose the standard deviations of parameter fluctuations to be tolerated *in vivo* are given by

$$\Delta S_{u,c_1} = 0.05 S_{u,c_1}, \quad \left\{ \Delta S_{u,c_i}, \Delta S_{l,c_i} \right\} = \left\{ 0.05 S_{u,c_i}, 0.05 S_{l,c_i} \right\}, i = 2,3$$

$$\Delta \gamma_{LacI} = 0.05 \gamma_{LacI}, \Delta \gamma_{TetR} = 0.05 \gamma_{TetR}$$

$$\Delta \gamma_{im,x_1} = 0.05 \gamma_{im,x_1}, \Delta \gamma_{m,x_1} = 0.05 \gamma_{m,x_1} \tag{7.30}$$

$$\Delta \gamma_{im,x_2} = 0.05 \gamma_{im,x_2}, \Delta \gamma_{m,x_2} = 0.05 \gamma_{m,x_2}$$

$$\Delta m_1 = 0.05 m_1, \Delta m_2 = 0.05 m_2, \Delta \mu = 0.05 \Delta \mu$$

and the environmental disturbances $v_i(t)$ are independent Gaussian noises with zero mean and unit variance.

5. Use GA to search for a set $\{c_3, I_{ATc}\}$ from corresponding libraries which minimizes the following cost function $J(c_3, I_{ATc})$

$$J_A\left(c_3, I_{ATc}\right) = \min_{c_3 \in Lib_{Tet}, I_2 \in [0.5,10]} E \int_{5152}^{12330} (y_{ss}(c_3, I_{ATc}, g_1) - y_d(g_1))^2 dg_1 \tag{7.31}$$

Then, the most adequate paring of c_3 and I_{ATc} are determined to be $\{c_3, I_{ATc}\} = \{T_3,\ 1.058\ \text{ng/ml}\}$. The Monte Carlo simulation results with 1000 runs of an amplifier are shown in Figure 7.15. Clearly, the I/O characteristics of the synthetic genetic transistor can match the desired response in a given input range $5152 \le x \le 12330$ (MEFL) with the intrinsic parameter fluctuations and environmental disturbances.

Table 7.1. The kinetic parameters of transistor dynamic model.

Kinetic parameter	Description	Value	Unit	Reference
γ_m	GFP degradation rate	0.35007×10^{-3}	min^{-1}	(Albano et al. 1996)
γ_{im}	Immature GFP degradation rate	5.7762×10^{-3}	min^{-1}	(Leveau and Lindow 2001)
m	Maturation rate	5.7762×10^{-3}	min^{-1}	(Leveau and Lindow 2001)
γ_{TetR}	TetR degradation rate	0.1386	min^{-1}	(Tuttle et al. 2005)
γ_{LacI}	LacI degradation rate	0.1386	min^{-1}	(Tuttle et al. 2005)
μ	Dilution rate due to cell growth	0.011946	min^{-1}	*

* The values are identified from the experiments.

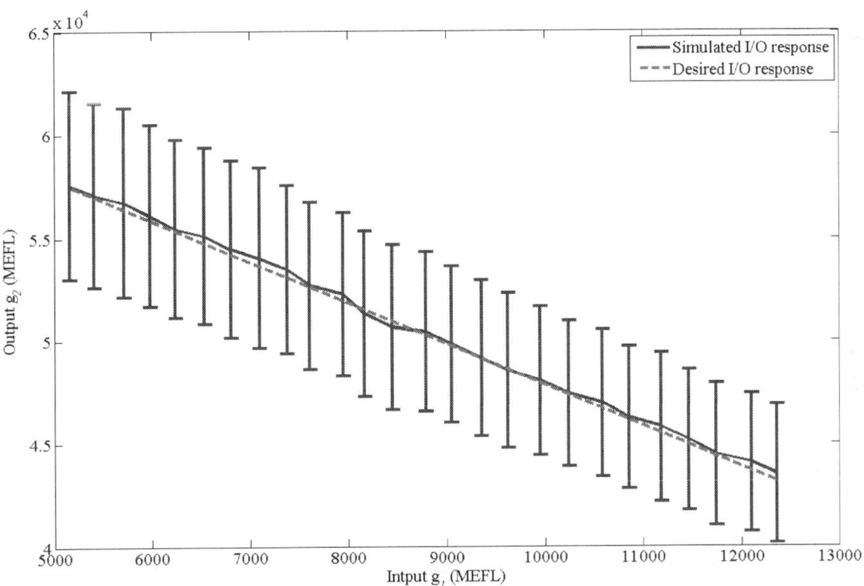

Figure 7.15. Monte Carlo simulation results of amplifier I/O response design example of synthetic genetic transistor. For amplifier design example of synthetic genetic transistor in Figure 7.14, the most adequate promoter-RBS component c_3 and ATc concentration I_{ATc} are searched as $\{c_3, I_{ATc}\} = \{T_3, 1.058 \text{ ng/ml}\}$ from the corresponding promoter-RBS library Lib_{TetR} and concentration range of inducer I_{ATc}. The Monte Carlo simulations are with 1000 runs. The blue line is the simulation result, and the error bars are the standard deviations. The red dash line is the desired I/O response generated by (7.29).

Color image of this figure appears in the color plate section at the end of the book.

7.2.5 Switch design example of synthetic genetic transistor

Consider the switch design example of the synthetic genetic transistor as shown in Figure 7.14B. The switch design procedure is similar to the amplifier design procedure of a synthetic genetic transistor. Firstly, to obtain the complete I/O characteristics of switching, promoter-RBS components $\{c_1,c_2\}=\{J_1,L_1\}$ are selected to obtain the maximum input operation range. The dynamic model and the steady state model have been described in (7.26) and (7.27). We can then provide the input operation range and desired I/O response as follows:

$$\text{Input operation range}:\ I_1 \in [0,10^5]\ (\text{nM}) \Rightarrow g_1 \in [181.5,150400]\ (\text{MEFL}) \quad (7.32)$$

and

$$y_d{}'(g_1)=L_s+\frac{(36000-L_s)}{1+(g_1/4000)^2} \tag{7.33}$$

where L_s denotes the low level of switching or basal level of promoter-RBS component c_3. Note that the standard deviations of parameter fluctuations that are supposed to be tolerated *in vivo* and from environmental disturbances are the same as in (7.30).

Finally, GA is employed to search a set $\{c_3,I_{ATc}\}$ from corresponding libraries to minimize the following cost function:

$$J_S(c_3,I_{ATc})=\min_{c_3\in Lib_{Tet}\ ,\ I_2\in[10^{-3},1]}\ \mathrm{E}\int_{181.5}^{150400}(y_{ss}(c_3,I_{ATc},g_1)-y_d{}'(g_1))^2 dg_1 \tag{7.34}$$

Then, the most adequate promoter-RBS component and ATc concentrations are found to be $\{c_3,I_{ATc}\}=\{T_1,0.182\ \text{ng/ml}\}$. Monte Carlo simulation results with 1000 runs of switch design of the synthetic genetic transistor circuit are shown in Figure 7.16. Clearly, similarly to the results for the amplification characteristics, the switching I/O characteristics of synthetic genetic transistors can match the desired I/O response under the intrinsic parameter fluctuations and environmental disturbances.

7.2.6 Summary

According to the above examples, the amplification or switching I/O characteristics of a synthetic genetic transistor with different design specifications can be achieved by selecting the most adequate promoter-RBS component c_3 and ATc concentration I_{ATc} using the proposed library-based searching method. However, not just the promoter-RBS component $c_3 \in Lib_{TetR}$ can be selected to achieve the amplification or switching design specification of the synthetic genetic transistor circuit, but also other

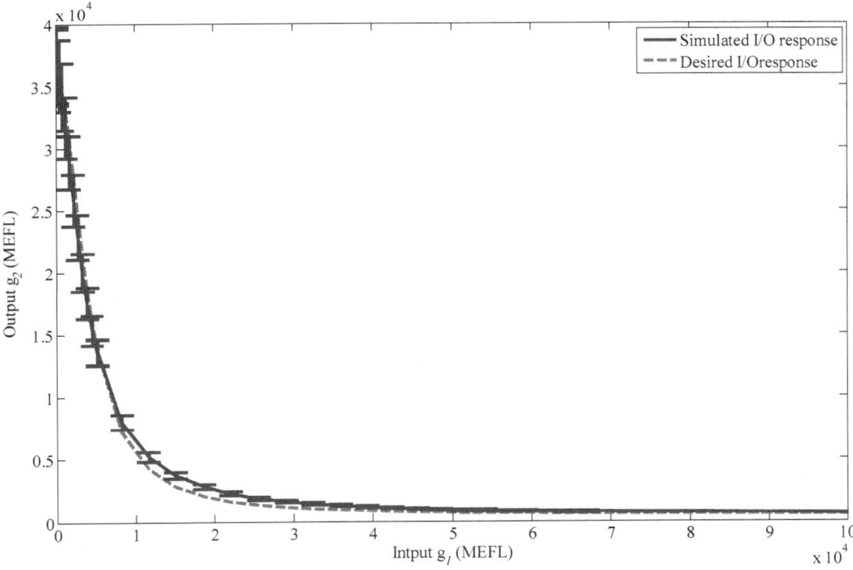

Figure 7.16. Monte Carlo simulations of switch design example of synthetic genetic transistor. For switch design example of synthetic genetic transistor in Figure 7.14, the most adequate promoter-RBS component c_3 and ATc concentration I_{ATc} are searched as $\{c_3, I_{ATc}\} = \{T_1, 0.182 \text{ ng/ml}\}$ from the corresponding promoter-RBS library Lib_{TetR} and concentration range of inducer I_{ATc}. The Monte Carlo simulations are with 1000 runs. The blue line is the simulation result, and the error bars are the standard deviations. The red dash line is the desired I/O response generated by (7.33).

Color image of this figure appears in the color plate section at the end of the book.

promoter-RBS components, i.e., Lib_{LacI} and $Lib_{Bad/araC}$, can be selected to achieve the desired I/O response. However, for various design specifications, more promoter-RBS libraries are needed to achieve these design specifications.

For the convenience of synthetic genetic transistor design for synthetic biologists, one look-up table for GFP has been built for the various design specifications as shown in Table 7.2. Based on various amplification gains in some specific operation range, the synthetic genetic transistors can be designed by first checking the look-up table. In future, more promoter-RBS components and inducer concentrations for different I/O characteristics of synthetic genetic transistors can be accumulated to build much larger look-up tables to match a lot of design specifications. From this look-up table, based on the desired design specifications, we can select the corresponding promoter-RBS components and inducer concentrations to synthesize the genetic transistors with desired I/O responses. Thus, less time will be spent on the design procedure as a designer will be able to easily construct transistors with the desired I/O characteristics.

Table 7.2. The look-up table with different gain specifications for synthetic genetic transistors.

Amplifier gain specifications of synthetic genetic transistor				
	Input range (MEFL)	Gain	Promoter-RBS component	Inducer concentration
Lib_{TetR}	2155~6495	−0.5	T_2	0.522 ng/ml
		−2	T_1	0.584 ng/ml
		−4	T_3	0.510 ng/ml
	8616~1.386×10^4	−2	T_3	1.016 ng/ml
Lib_{LacI}	1218~4208	−5	L_3	765 nM
		−10	L_1	489 nM
	4715~1.048×10^4	−5	L_1	983 nM
	1.449×10^4~1.902×10^4	−2	L_3	432 nM

References

Albano, C.R., Randers Eichhorn, L., Chang, Q., Bentley, W.E. and Rao, G. 1996. Quantitative measurement of green fluorescent protein expression. Biotechnology Techniques 10: 953–958.

Chen, B.S. and Wu, C.H. 2009. A systematic design method for robust synthetic biology to satisfy design specifications. BMC Syst Biol 3: 66.

Chen, B.S., Chang, C.H., Wang, Y.C., Wu, C.H. and Lee, H.C. 2011. Robust model matching design methodology for a stochastic synthetic gene network. Math Biosci 230: 23–36.

Lee, Y.Y., Hsu, C.Y., Lin, L.J, Chang, C.C., Cheng, H.C., Yeh, T.H., Hu, R.H., Lin, C., Xie, Z. and Chen, B.S. 2013. Systematic design methodology for robust genetic transistors based on I/O specifications via promoter-RBS libraries. BMC Syst Biol 7: 109.

Leveau, J.H. and Lindow, S.E. 2001. Predictive and interpretive simulation of green fluorescent protein expression in reporter bacteria. J Bacteriol 183: 6752–6762.

Neamen, D.A. 2007. Microelectronics : circuit analysis and design. McGraw-Hill, New York.

Tuttle, L.M., Salis, H., Tomshine, J. and Kaznessis, Y.N. 2005. Model-Driven Designs of an Oscillating Gene Network. Biophysical Journal 89: 3873–3883.

Varadarajan, P.A. and Del Vecchio, D. 2009. Design and Characterization of a Three-terminal Transcriptional Device Through Polymerase Per Second. IEEE Transactions on Nanobioscience 8: 281–289.

Wu, C.H., Lee, H.C. and Chen, B.S. 2011. Robust synthetic gene network design via library-based search method. Bioinformatics 27: 2700–2706.

8

Communication and Synchronization of a Population of Coupled Synthetic Gene Networks

Recently, a simple synthetic device was engineered in a cell, and several cells were then combined, so that their connections allowed the construction of a more complex synthetic gene network, i.e., so-called multi-cellular engineered gene networks. This approach not only uses cellular consortia as an efficient way of engineering complex gene networks, but also demonstrates the great potential for reutilization of small parts of the gene network. Therefore, communication and synchronization of a population of synthetic gene networks is an important design in practical applications, because such a population of synthetic gene networks distributed over different host cells need to exploit molecular phenomena simultaneously in order to emerge a biological phenomenon. However, this intercellular communication and synchronization may be corrupted by intrinsic kinetic parameter fluctuations and extrinsic environmental noise. Therefore, robust synchronization is important in the design of nonlinear stochastic coupled synthetic gene networks with intrinsic kinetic parameter fluctuations and extrinsic molecular noises.

In this chapter, the robust synchronization of a population of synthetic genetic oscillators is given as a design example. Initially, the condition for robust synchronization of coupled synthetic genetic oscillators was derived based on Hamilton-Jacobi inequality (HJI). We found that if the synchronization robustness can confer enough intrinsic robustness to tolerate intrinsic parameter fluctuation and extrinsic robustness to filter the environmental noise, then robust synchronization of coupled synthetic genetic oscillators is guaranteed. If the synchronization robustness of a population of nonlinear stochastic coupled synthetic genetic oscillators distributed over different host cells could not be maintained, then robust synchronization could be enhanced by external control input through the communication of quorum sensing molecules. In order to simplify the analysis and design of robust synchronization of nonlinear stochastic coupled synthetic genetic oscillators, the fuzzy interpolation method was employed to interpolate several local linear stochastic coupled systems to approximate the nonlinear stochastic coupled networks so that the HJI-based synchronization design problem could be replaced by a simple linear matrix inequality (LMI)-based design problem, which could be solved with the help of LMI toolbox in MATLAB easily. If the synchronization robustness criterion is violated, external control scheme by adding inducer can be designed to improve synchronization robustness of coupled synthetic genetic oscillators. The investigated robust synchronization criteria and proposed external control method are useful for a population of coupled synthetic gene networks with emergent synchronization behavior, especially for multi-cellular complex synthetic gene network in the future.

8.1 Nonlinear Stochastic Coupled Synthetic Genetic Oscillators under Intrinsic Kinetic Parameter Fluctuations and External Molecular Noises

8.1.1 Model description

Before discussion of synchronization of more general synthetic genetic oscillators, we here provide a design example of coupled repressilators to illustrate the interesting phenomenon of synchronization in coupled dynamic cells. Then model description, definition, and theoretical results of robust synchronization of more general coupled synthetic genetic networks will be introduced in the sequel. The repressilator is a network of three genes, the products of which inhibit the transcription of each other in a cyclical manner (Garcia-Ojalvo et al. 2004). The gene *lacI* (from *E. coli*) codes for the protein LacI, which inhibits the transcription of the gene *tetR*. The product of the latter, TetR, inhibits the transcription of gene *cI* (from λ phage); the protein product CI in turn inhibits the expression of *lacI*, thus completing the cycle. Garcia-Ojalvo et al. proposed a modular addition to the repressilator, with the aim of coupling a population of cells containing this network (Garcia-Ojalvo et al. 2004). The basic mechanism of communication among the cells is based on quorum sensing, which was first discovered in the bioluminescent bacteria, *Vibrio fisceri*. These bacteria exhibit collective behaviors, mainly using two proteins. The first protein, LuxI, synthesizes

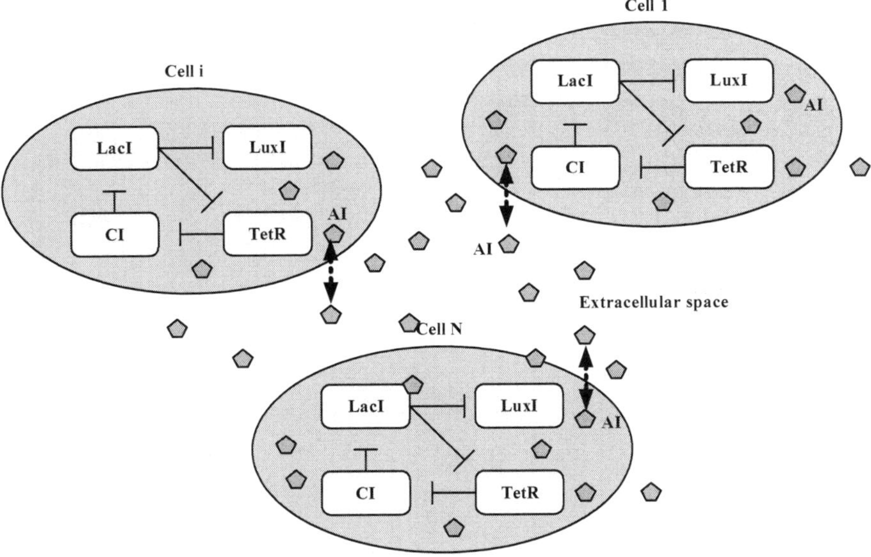

Figure 8.1. Synchronization scheme of N coupled genetic oscillators distributed in different host cells by the quorum sensing mechanism.

a small molecule known as an auto-inducer (AI), which diffuses freely through the cell membrane. The second protein, LuxR, binds to the AI molecule to form a complex, which subsequently activates transcription of various genes (a detailed description of this molecular pathway is provided by Garcia-Ojalvo et al. 2004). Several studies have indicated that quorum sensing can be engineered and used as an intercellular signaling system in *E. coli* (Garcia-Ojalvo et al. 2004, Wang and Chen 2005, Wang et al. 2006).

Based on the synchronization scheme of coupled repressilators via quorum sensing mechanism (Figure 8.1), the mRNA dynamics in the cell $i = 1, 2, \ldots, N$ are governed by repressible transcription of all three genes of the repressilator plus transcription activation of the additional copy of the *lacI* gene, and by mRNA degradation (Wang and Chen 2005).

$$\frac{dx_{ai}(t)}{dt} = -\gamma_m x_{ai}(t) + \frac{\alpha_a}{\mu + x_{Ci}^n(t)}$$

$$\frac{dx_{bi}(t)}{dt} = -\gamma_m x_{bi}(t) + \frac{\alpha_b}{\mu + x_{Ai}^n(t)} \qquad , \ i = 1, 2, \ldots, N \qquad (8.1)$$

$$\frac{dx_{ci}(t)}{dt} = -\gamma_m x_{ci}(t) + \frac{\alpha_c}{\mu + x_{Bi}^n(t)} + \frac{\alpha_S x_{Si}(t)}{\mu_S + x_{Si}(t)}$$

where x_{ai}, x_{bi} and x_{ci} are the concentration of mRNA transcribed from *tetR*, *cI* and *lacI* in cell i, respectively; concentrations of the corresponding proteins are represented by x_{Ai}, x_{Bi} and x_{Ci}, respectively. The concentration of AI inside each cell is denoted by x_{Si}. α_a, α_b, and α_c are the dimensionless transcription rate in the absence of the repressor, and μ is the repression coefficient. α_s is the maximal contribution to *lacI* transcription in the presence of saturating amounts of AI, and μ_s is the activation coefficient. γ_m is the respective dimensionless degradation rate of mRNA for *tetR*, *cI* and *lacI* in the cell, and n is the Hill coefficient.

The dynamics of proteins TetR, CI and LacI are given respectively as (Wang and Chen 2005)

$$\frac{dx_{Ai}(t)}{dt} = -\gamma_p x_{Ai}(t) + \beta_A x_{ai}(t)$$

$$\frac{dx_{Bi}(t)}{dt} = -\gamma_p x_{Bi}(t) + \beta_B x_{bi}(t) \, , \, i = 1, 2, \ldots, N \tag{8.2}$$

$$\frac{dx_{Ci}(t)}{dt} = -\gamma_p x_{Ci}(t) + \beta_C x_{ci}(t)$$

where parameters β_A, β_B and β_C are the translation rates of the proteins from their mRNA, and γ_p represents the dimensionless degradation rate of proteins TetR, CI, and LacI in the cell. The intercellular protein AI in cell i is synthesized by the protein LuxI and diffuses through the cell wall where it undergoes degradation, leading to the following equation (Chen and Hsu 2012).

$$\frac{dx_{Si}(t)}{dt} = -\gamma_s x_{Si}(t) + \beta_s x_{Ai}(t) - \eta_s Q_e \left(x_{Si}(t) - N^{-1} \sum_{j=1}^{N} x_{Sj}(t) \right), \, 0 \le Q_e \le 1 \tag{8.3}$$

where η_s measures the diffusion rate of AI across the cell membrane, β_s is the synthesis rate of AI, and γ_s gives the rate of decay of AI. Consequently, the whole coupled synthetic system is expressed by (8.1)–(8.3), which can be represented by the following more generalized nonlinear dynamic equation (Chen and Hsu 2012)

$$\dot{x}_i(t) = f(x_i(t)) + \sum_{j=1}^{N} c_{ij} g(x_j(t)), \, i = 1, 2, \ldots, N \tag{8.4}$$

where $x_i(t) = \left(x_{ai}(t), x_{bi}(t), x_{ci}(t), x_{Ai}(t), x_{Bi}(t), x_{Ci}(t), x_{Si}(t) \right)^T \in R^m$ is the state vector of the ith synthetic oscillator; $f(\cdot): R^m \to R^m$ is a smooth nonlinear function that characterizes the behavior of the synthetic oscillator; $g(\cdot): R^m \to R^m$ is a smooth nonlinear inner-coupling function; and $C = \left(c_{ij} \right)_{N \times N} \in R^{N \times N}$ is the coupling configuration matrix, where $c_{ij} > 0$

means that the ith synthetic oscillator is coupled with the jth synthetic oscillator directly, otherwise $c_{ij} = 0$. Assume that the diagonal elements of C satisfy $c_{ii} = -\sum\limits_{j=1, j\neq i}^{N} c_{ij}$.

8.1.2 Synthetic genetic oscillators under intrinsic parameter fluctuations

In general, the synthetic genetic oscillators suffer kinetic parameter fluctuations from transcription control, alternative splicing, translation, genetic mutation and diffusion to biological modification of transcription factors. In this situation, dynamic equations of coupled synthetic genetic oscillators in (8.1)–(8.3) are modified as

$$\frac{dx_{ai}(t)}{dt} = -\left(\gamma_m + \Delta\gamma_m n_i(t)\right) x_{ai}(t) + \frac{\left(\alpha_a + \Delta\alpha_a n_i(t)\right)}{\mu + x_{Ci}^n(t)}$$

$$\frac{dx_{bi}(t)}{dt} = -\left(\gamma_m + \Delta\gamma_m n_i(t)\right) x_{bi}(t) + \frac{\left(\alpha_b + \Delta\alpha_b n_i(t)\right)}{\mu + x_{Ai}^n(t)}$$

$$\frac{dx_{ci}(t)}{dt} = -\left(\gamma_m + \Delta\gamma_m n_i(t)\right) x_{ci}(t) + \frac{\left(\alpha_c + \Delta\alpha_c n_i(t)\right)}{\mu + x_{Bi}^n(t)} + \frac{\left(\alpha_S + \Delta\alpha_S n_i(t)\right)x_{Si}(t)}{\mu_S + x_{Si}(t)}$$

$$\frac{dx_{Ai}(t)}{dt} = -\left(\gamma_p + \Delta\gamma_p n_i(t)\right) x_{Ai}(t) + \left(\beta_A + \Delta\beta_A n_i(t)\right)x_{ai}(t)$$

$$\frac{dx_{Bi}(t)}{dt} = -\left(\gamma_p + \Delta\gamma_p n_i(t)\right) x_{Bi}(t) + \left(\beta_B + \Delta\beta_B n_i(t)\right)x_{bi}(t) \qquad (8.5)$$

$$\frac{dx_{Ci}(t)}{dt} = -\left(\gamma_p + \Delta\gamma_p n_i(t)\right)x_{Ci}(t) + \left(\beta_C + \Delta\beta_C n_i(t)\right)x_{ci}(t)$$

$$\frac{dx_{Si}(t)}{dt} = -\left(\gamma_s + \Delta\gamma_s n_i(t)\right)x_{Si}(t) + \left(\beta_s + \Delta\beta_s n_i(t)\right)x_{Ai}(t)$$

$$- \left(\eta_s + \Delta\eta_s n_i(t)\right)Q_e\left(x_{Si}(t) - N^{-1}\sum\limits_{j=1}^{N} x_{Sj}(t)\right)$$

where $\Delta\alpha_j, j \in \{a,b,c,S\}$, $\Delta\beta_j, j \in \{A,B,C,s\}$, $\Delta\gamma_j, j \in \{m,p,s\}$ and $\Delta\eta_s$ denote the amplitudes of the kinetic parameter fluctuations and $n_i(t)$ is random white noise with zero-mean and unit variance, i.e., $\Delta\alpha_j$, $\Delta\beta_j$, $\Delta\gamma_j$ and $\Delta\eta_s$ denote the deterministic part of the stochastic kinetic parameter fluctuations $\Delta\alpha_j n_i(t)$, $\Delta\beta_j n_i(t)$, $\Delta\gamma_j n_i(t)$ and $\Delta\eta_s n_i(t)$, respectively, and $n_i(t)$ absorbs the stochastic property of random intrinsic kinetic parameter fluctuations. The covariances of the stochastic parameter fluctuations $\Delta\alpha_j n_i(t)$, $\Delta\beta_j n_i(t)$, $\Delta\gamma_j n_i(t)$ and $\Delta\eta_s n_i(t)$ are given as $cov\left(\xi n_i(t), \xi n_i(t)\right) = \xi^2 \delta_{t,\tau}$ with $\xi \in \{\Delta\alpha_j, \Delta\beta_j, \Delta\gamma_j, \Delta\eta_s\}$, respectively, where $\delta_{t,\tau}$ denotes the delta function, that is, $\delta_{t,\tau} = 1$ if $t = \tau$

and $\delta_{t,\tau} = 0$ if $t \neq \tau$, i.e., $\Delta\alpha_j$, $\Delta\beta_j$, $\Delta\gamma_j$ and $\Delta\eta_s$ denote the corresponding standard deviations of the stochastic parameter fluctuations $\Delta\alpha_j n_i(t)$, $\Delta\beta_j n_i(t)$, $\Delta\gamma_j n_i(t)$ and $\Delta\eta_s n_i(t)$, respectively. Thus, the nonlinear stochastic coupled synthetic genetic oscillators under intrinsic kinetic parameter fluctuations in the host cell i can be represented by (Chen and Hsu 2012)

$$\dot{x}_i(t) = \left(f(x_i(t)) + \sum_{j=1}^{N} c_{ij} g(x_j(t)) \right) + \left(f_W(x_i(t)) + \sum_{j=1}^{N} c_{ij} g_W(x_j(t)) \right) n_i(t) \quad (8.6)$$

where $\left(f_W(x_i(t)) + \sum_{j=1}^{N} c_{ij} g_W(x_j(t)) \right) n_i(t)$ denotes the intrinsic kinetic parameter fluctuations of synthetic genetic oscillator in the host cell i. For the convenience of analysis and control design of the synchronization in synthetic genetic oscillators inserted into different host cells, the nonlinear stochastic equation in (8.6) can be represented by the following Ito's stochastic differential equation

$$dx_i(t) = \left(f(x_i(t)) + \sum_{j=1}^{N} c_{ij} g(x_j(t)) \right) dt + \left(f_W(x_i(t)) + \sum_{j=1}^{N} c_{ij} g_W(x_j(t)) \right) dw_i(t) \quad (8.7)$$

where $w_i(t)$ is a standard Wiener process or Brownian motion with $dw_i(t) = n_i(t)dt$ to represent the random kinetic fluctuations of synthetic gene circuits.

8.1.3 Synthetic genetic oscillators under external environmental molecular noise

In general, a synthetic genetic oscillator *in vivo* also suffers from extrinsic environmental molecular noise, such as transmitted noise from upstream and global noise affecting all cells. Therefore, the nonlinear stochastic equation of coupled oscillators in (8.7) should be modified to mimic realistic dynamic behavior, as follows:

$$dx_i(t) = \left(f(x_i(t)) + \sum_{j=1}^{N} c_{ij} g(x_j(t)) + h_i v_i(t) \right) dt + \left(f_W(x_i(t)) + \sum_{j=1}^{N} c_{ij} g_W(x_j(t)) \right) dw_i(t) \quad (8.8)$$

where the signal vector $v_i(t)$ denotes extrinsic environmental molecular noise from the environment and h_i denotes the noise-coupling matrix in cell i. For the nonlinear stochastic dynamic of the ith oscillator in (8.8), $f(x_i(t))$ denotes the characteristic of synthetic oscillator, $\sum_{j=1}^{N} c_{ij} g(x_j(t))$ denotes the coupling (intercellular) signals from other oscillators, $h_i v_i(t)$

denotes the environmental (extracellular) noises, and the last term $\left(f_W(x_i(t)) + \sum_{j=1}^{N} c_{ij} g_W(x_j(t)) \right) dw_i(t)$ is the random intrinsic fluctuations (noises).

8.2 Robust Synchronization Control Design for Nonlinear Synthetic Genetic Oscillators under Intrinsic Kinetic Parameter Fluctuations and External Environmental Molecular Noise

The nonlinear stochastic coupled synthetic genetic oscillators in (8.8) are said to reach synchronization asymptotically if $x_1(t) = x_2(t) = \cdots = x_N(t) = s(t)$ in probability as $t \to t_f$ or

$$E\left(x_i(t) - s(t)\right) = 0 \text{ as } t \to t_f \tag{8.9}$$

in which $s(t) \in R^m$ is the synchronization solution satisfying

$$ds(t) = \left(f(s(t)) + \sum_{j=1}^{N} c_{ij} g(s(t)) + h_s v_i(t) \right) dt + \left(f_W(s(t)) + \sum_{j=1}^{N} c_{ij} g_W(s(t)) \right) dw_i(t) \tag{8.10}$$

Let us denote the synchronization error signal for synthetic genetic oscillators as

$$e_i(t) = x_i(t) - s(t), \ i = 1, 2, \ldots, N \tag{8.11}$$

According to (8.8), the synchronization error dynamics for cell i are then described by

$$de_i(t) = \left(f(x_i(t)) - f(s(t)) + \sum_{j=1}^{N} c_{ij} \left(g(x_j(t)) - g(s(t)) \right) + \left(h_i - h_s \right) v_i(t) \right) dt$$

$$+ \left(f_W(x_i(t)) - f_W(s(t)) + \sum_{j=1}^{N} c_{ij} \left(g_W(x_j(t)) - g_W(s(t)) \right) \right) dw_i(t) \tag{8.12}$$

for $i = 1, 2, \ldots, N$ which can be augmented as

$$de = \left(F(x,s) + (C \otimes I_m)G(x,s) + Hv \right) dt + \left(F_W(x,s) + (C \otimes I_m)G_W(x,s) \right) dw \tag{8.13}$$

where

$$x = \begin{pmatrix} x_1 \\ \vdots \\ x_N \end{pmatrix}, e = \begin{pmatrix} e_1 \\ \vdots \\ e_N \end{pmatrix}, v = \begin{pmatrix} v_1 \\ \vdots \\ v_N \end{pmatrix}, w = \begin{pmatrix} w_1 \\ \vdots \\ w_N \end{pmatrix}, H = \begin{pmatrix} h_1 - h_s & & 0 \\ & \ddots & \\ 0 & & h_N - h_s \end{pmatrix},$$

and

$$F(x,s) = \begin{pmatrix} f(x_1(t)) - f(s(t)) \\ \vdots \\ f(x_N(t)) - f(s(t)) \end{pmatrix}, G(x,s) = \begin{pmatrix} g(x_1(t)) - g(s(t)) \\ \vdots \\ g(x_N(t)) - g(s(t)) \end{pmatrix},$$

$$F_W(x,s) = \begin{pmatrix} f_W(x_1(t)) - f_W(s(t)) & & 0 \\ & \ddots & \\ 0 & & f_W(x_N(t)) - f_W(s(t)) \end{pmatrix},$$

$$G_W(x,s) = \begin{pmatrix} g_W(x_1(t)) - g_W(s(t)) & & 0 \\ & \ddots & \\ 0 & & g_W(x_N(t)) - g_W(s(t)) \end{pmatrix}$$

Suppose the influence of extrinsic environmental molecular noise on the synchronization error can be bounded by the following noise-filtering level ρ of coupled synthetic oscillators (Chen and Zhang 2004, Chen and Wang 2006, Chen and Wu 2008)

$$\frac{E\int_0^{t_f} e^T(t)R e(t)dt}{E\int_0^{t_f} v^T(t)v(t)dt} \le \rho^2 \quad \text{or} \quad E\int_0^{t_f} e(t)^T Re(t)dt \le \rho^2 E\int_0^{t_f} v(t)^T v(t)dt \quad (8.14)$$

for all possible environmental noises $v(t)$, where ρ^2 denotes the upper bound of the effect of $v(t)$ on synchronization error from the mean energy point of view, i.e., the noise filtering level ρ in (8.14) denotes the upper bound of the noise-filtering ability ρ_0 of coupled synthetic genetic oscillators, and R is a symmetric weighting matrix to be specified by designer.

Remark 8.1

i) The inequality in (8.14) means that the effect of extrinsic environmental molecular noise on the synchronization error is less than ρ from the mean energy point of view, i.e., the value of the noise-filtering ability ρ_0 is the lower bound of ρ. Because the statistics of extrinsic molecular noise may be unavailable or uncertain, it is very difficult to obtain the noise filtering ability ρ_0 for all possible extrinsic noises $v(t)$ directly and only the upper bound ρ of the noise-filtering ability ρ_0 can be given in (8.14) at first. Then, we will decrease the upper bound ρ to as small a value as possible to approach its lower bound for the noise-filtering ability ρ_0, i.e., to get ρ_0 by minimizing ρ (or $\rho_0 = \min \rho$) indirectly. If the noise filtering ability ρ_0 is small, it means that the environmental molecular noise has less influence on synchronization and vice versa. It will be further discussed in the sequel.

ii) If the extrinsic environmental molecular noise v is deterministic, then the expectation on $v(t)$ in (8.14) should be neglected. If the initial condition $e(0)$ is considered, then the noise-filtering level in (8.14) should be modified as follows:

$$E\int_0^{t_f} e(t)^T Re(t)dt \le EV(e(0)) + \rho^2 E\int_0^{t_f} v(t)^T v(t)dt \tag{8.15}$$

for some Lyapunov function $V\left(e(0)\right)$, i.e., the energy due to the initial condition $e(0)$ should be considered in the influence of noise on synchronization (Zhang et al. 2005).

8.3 Synchronization Robustness Criterion of Coupled Synthetic Genetic Oscillators

Based on the synchronization error dynamic equation in (8.13) and the H_∞ noise filtering performance in (8.14) or (8.15), we obtain the following robust synchronization result for nonlinear stochastic coupled synthetic genetic oscillators under intrinsic kinetic parameter fluctuations and extrinsic environmental molecular noise.

Proposition 8.1: If there exists a positive function $V(e) > 0$ with $V(0) = 0$ solving the following HJI,

$$e^T Re + \left(\frac{\partial V(e)}{\partial e}\right)^T \left(F(x,s) + (C \otimes I_m)G(x,s)\right) + \frac{1}{4\rho^2}\left(\frac{\partial V(e)}{\partial e}\right)^T HH^T \left(\frac{\partial V(e)}{\partial e}\right)$$
$$+ \frac{1}{2}\left(F_W(x,s) + (C \otimes I_m)G_W(x,s)\right)\frac{\partial^2 V(e)}{\partial e^2}\left(F_W(x,s) + (C \otimes I_m)G_W(x,s)\right) < 0 \tag{8.16}$$

then the stochastic intrinsic noise can be tolerated (i.e., the synchronization of coupled synthetic genetic networks can not be destroyed by intrinsic noise) and the influence of extrinsic environmental molecular noise $v(t)$ on the synchronization of the nonlinear stochastic coupled synthetic oscillation systems in (8.8) is less than or equal to a prescribed filtering level ρ, i.e., the inequality in (8.14) holds.

Proof: See Appendix 8.1

Since ρ denotes an upper bound of the effect of $v(t)$ on synchronization, the real effect can be obtained by minimizing ρ to as small a value as possible. Therefore, the noise-filtering ability of synchronized oscillators on $v(t)$ can be obtained by solving the following constrained optimization:

$$\rho_0 = \min \rho$$
$$\text{subject to HJI in (8.16) with } V(e) > 0 \tag{8.17}$$

i.e., the noise-filtering ability ρ_0 on $v(t)$ for the synchronized synthetic oscillators can be evaluated by solving the constrained optimization in (8.17). The noise-filtering ability ρ_0 of the synchronized synthetic oscillators in (8.17) can be obtained by decreasing ρ until no positive solution $V(e) > 0$ exists for HJI in (8.16) again.

If the noise-filtering ability ρ_0 can not satisfy the designer's specification, in order to enhance the noise filtering of extrinsic noise, we need to specify the design parameters of nonlinear stochastic coupled synthetic genetic oscillators, for example, the kinetic parameters α_a, α_b, α_c, β_A, β_B, β_C, γ_m, and γ_p in (8.5) to solve the constrained optimization in (8.17) to enhance the noise-filtering ability, i.e.,

$$\rho_0 = \min_{\alpha_a, \alpha_b, \alpha_c, \beta_A, \beta_B, \beta_C, \gamma_m, \gamma_p} \rho$$
$$\text{subject to HJI in (8.16) with } V(e) > 0 \tag{8.18}$$

Before the discussion on the synchronization robustness criterion of coupled synthetic genetic oscillators, some definitions on synchronization robustness, intrinsic robustness and extrinsic robustness are given as follows (Chen and Hsu 2012):

1) Synchronization robustness: the ability of coupled synthetic genetic oscillators to resist both intrinsic noise and extrinsic noise so that the synchronization can be maintained.
2) Intrinsic robustness: the ability of coupled synthetic genetic oscillator to tolerate intrinsic parameter fluctuation to maintain synchronization.
3) Extrinsic robustness: the filtering ability to attenuate the effect of environmental noise on the synchronization of coupled synthetic genetic network.

Remark 8.2

Substituting the noise-filtering ability ρ_0 of (8.17) into (8.16) in Proposition 8.1, we get the following equivalent synchronization robustness criterion

$$\underbrace{\frac{1}{2}\Big(F_W(x,s)+(C\otimes I_m)G_W(x,s)\Big)\frac{\partial^2 V(e)}{\partial e^2}\Big(F_W(x,s)+(C\otimes I_m)G_W(x,s)\Big)}_{\text{intrinsic robustness}}$$

$$\underbrace{+e^T Re+\frac{1}{4\rho_0^2}\left(\frac{\partial V(e)}{\partial e}\right)^T HH^T\left(\frac{\partial V(e)}{\partial e}\right)}_{\text{extrinsic robustness}}\le \underbrace{-\left(\frac{\partial V(e)}{\partial e}\right)^T\Big(F(x,s)+(C\otimes I_m)G(x,s)\Big)}_{\text{synchronization robustness}} \quad (8.19)$$

The first term on the left hand side of (8.19) indicates the intrinsic robustness to tolerate the intrinsic parameter fluctuation in (8.13) because this term is induced by intrinsic noise (or random parameter fluctuation), the second and third term on the left hand side are due to the noise filtering in (8.14) and indicate the extrinsic robustness to filter the extrinsic noise with the noise filtering ability ρ_0, and the term on the right hand side of (8.19) indicates the synchronization robustness of the coupled synthetic gene networks. The biological meaning of synchronization robustness criterion in (8.19) is that if the synchronization robustness can confer both the intrinsic robustness to tolerate intrinsic parameter fluctuation and extrinsic robustness to filter the environmental noise, then the coupled synthetic networks will synchronize with a noise filtering ability ρ_0. If the synchronization robustness criterion in (8.19) is violated, then the synchronization of coupled synthetic gene networks may not be achieved due to the intrinsic parameter fluctuation and extrinsic noise.

In general, it is still very difficult to solve the second-order HJI in (8.16) with $V(e) > 0$ and $V(0) = 0$ to guarantee robust synchronization of nonlinear stochastic coupled synthetic genetic oscillators with a prescribed attenuation level ρ under intrinsic kinetic parameters fluctuations and extrinsic environmental molecular noise or to solve the constrained minimization in (8.18) for robust synchronization design to achieve the optimal molecular noise filtering of the synchronized coupled synthetic oscillators. Recently, the fuzzy dynamic model has been widely used to interpolate several local dynamic models to efficiently approximate a nonlinear dynamic system (Takagi and Sugeno 1985, Chen et al. 1999). Hence, in this situation, we employ the T-S fuzzy model to interpolate several linear synthetic stochastic oscillators at different local operation points to efficiently and globally approximate the error dynamic in (8.13), so that the analysis and design procedure for robust synchronization of nonlinear stochastic coupled synthetic genetic oscillators can be simplified.

8.4 Robust Synchronization Design of Synthetic Genetic Oscillators via T-S Fuzzy Methodology

In this section, the T-S fuzzy method is employed to simplify the analysis and design procedure for robust synchronization of nonlinear stochastic coupled synthetic oscillators under intrinsic kinetic parameter fluctuations and extrinsic environmental molecular noise. The T-S fuzzy model for the synchronization error dynamics is described by fuzzy if-then rules. The kth rule of the fuzzy model for the synchronization error dynamics for cell i in (8.12) is proposed in the following form (Takagi and Sugeno 1985, Chen et al. 1999, Chen and Zhang 2004):

Rule k:

If $z_{1,i}(t)$ is F_{k1} and $z_{2,i}(t)$ is F_{k2} ... and $z_{g,i}(t)$ is F_{kg}, then

$$de_i = \left(A_k e_i + \sum_{j=1}^{N} c_{ij} B_k e_j + H_i v_i \right) dt + \left(A_{Wk} e_i + \sum_{j=1}^{N} c_{ij} B_{Wk} e_j \right) dw_i \qquad (8.20)$$

for $k = 1, 2, \cdots, L$, where $z_{g,i}$ is the element of premise variables of the ith coupled oscillation system, i.e., $z_i = [z_{1,i}, \ldots, z_{g,i}]^T$; F_{kg} is the fuzzy set; A_k, B_k, A_{Wk}, and B_{Wk} are the fuzzy system matrices; L is the number of if-then rules; and g is the number of premise variables. The physical meaning of fuzzy rule k is that if the premise variables $z_{1,i}(t), z_{2,i}(t), \cdots, z_{g,i}(t)$ are with the fuzzy sets $F_{k1}, F_{k2}, \cdots, F_{kg}$, then the synchronization error dynamics in (8.12) can be represented by interpolating the linearized synchronization error dynamics in (8.20) via the fuzzy basis. The fuzzy synchronization error dynamics in (8.20) is referred as follows (Chen and Hsu 2012)

$$de_i = \sum_{k=1}^{L} \mu_{k,i}(z_i) \left(\left(A_k e_i + \sum_{j=1}^{N} c_{ij} B_k e_j + H_i v_i \right) dt + \left(A_{Wk} e_i + \sum_{j=1}^{N} c_{ij} B_{Wk} e_j \right) dw_i \right) \qquad (8.21)$$

where $\mu_{k,i}(z_i) \triangleq \dfrac{\prod_{j=1}^{g} F_{kj}(z_{j,i})}{\sum_{k=1}^{L} \prod_{j=1}^{g} F_{kj}(z_{j,i})}$, $F_{kj}(z_{j,i})$ is the grade of membership of $z_{j,i}(t)$

in F_{kj} or the possibility function of $z_{j,i}(t)$ in F_{kj}, and $\mu_k(z_i)$ is called fuzzy basis function for $k = 1, 2, \ldots, L$. The denominator or $\sum_{k=1}^{L} \prod_{j=1}^{g} F_{kj}(z_{j,i})$ in the above fuzzy basis function is only for normalization, so that the total sum of fuzzy basis is $\sum_{k=1}^{L} \mu_{k,i}(z_i) = 1$. The physical meaning of (8.21) is that the fuzzy stochastic system interpolates L local linear stochastic systems through nonlinear basis $\mu_k(z_i)$ to approximate the nonlinear stochastic system in (8.13). In this situation, the nonlinear stochastic coupled oscillation systems in (8.13) can be represented by the fuzzy interpolation system as follows:

$$de = \left(F(x,s) + (C \otimes I_m)G(x,s) + Hv\right)dt + \left(F_W(x,s) + (C \otimes I_m)G_W(x,s)\right)dw$$

$$= \sum_{k=1}^{L} \mu_k(z)\left(\left((I_N \otimes A_k + C \otimes B_k)e + Hv\right)dt + \left((I_N \otimes A_{Wk} + C \otimes B_{Wk})e\right)dw\right) \tag{8.22}$$

where $\mu_k(z) = diag(\mu_{k,1}(z_1),...,\mu_{k,N}(z_N))$ and $z = [z_1,...,z_N]^T$.

Remark 8.3

Takagi and Sugeno have proposed the systematic method to build T-S fuzzy model for nonlinear function approximation by the system identification tool (Takagi and Sugeno 1985), i.e., the local system matrix A_k, B_k, A_{Wk}, and B_{Wk} in (8.21) or (8.22) can be identified by least square estimation method. On the other hand, many studies have proved that the T-S fuzzy model can approximate a continuous function with any degree of accuracy. Actually, there is still some fuzzy approximation error in (8.22). In the robust synchronization control design, for simplicity, the fuzzy approximation error can be merged into the external noise, which could be efficiently attenuated by the proposed H$_\infty$ robust synchronization control design in the sequel.

After investigating the approximation of nonlinear stochastic coupled synthetic oscillators by the fuzzy interpolation method, in order to avoid solving the nonlinear constrained optimization problem in (8.18) for the robust synchronization design problem of coupled synthetic oscillators under intrinsic kinetic parameter fluctuation and extrinsic environmental molecular noise, the measurement procedure for the noise-filtering ability of synchronized synthetic genetic oscillators could also be simplified by the fuzzy approximation method. Then, we get the following result.

Proposition 8.2: If there exists a positive definite symmetric matrix $P > 0$ solving the following LMIs,

$$\begin{bmatrix} R + P(I_N \otimes A_k + C \otimes B_k) + (I_N \otimes A_k + C \otimes B_k)^T P & \\ +(I_N \otimes A_{Wk} + C \otimes B_{Wk})^T P(I_N \otimes A_{Wk} + C \otimes B_{Wk}) & PH \\ H^T P & -\rho^2 I \end{bmatrix} < 0 \tag{8.23}$$

for $k = 1, 2, ..., L$, then the noise-filtering level ρ in (8.14) holds, or the intrinsic kinetic parameter fluctuations are tolerated by the synchronized synthetic genetic oscillators, and the influence of extrinsic environmental molecular noise $v(t)$ on the synchronized synthetic oscillation systems in (8.13) is less than or equal to a prescribed filtering level ρ.

Proof: See Appendix 8.2

Therefore, the optimal noise-filtering design of synchronized oscillation systems obtained by solving the HJI-constrained optimization problem in

(8.18) could be replaced by solving the following constrained optimizations, respectively

$$\rho_0 = \min_{\alpha_a,\alpha_b,\alpha_c,\beta_A,\beta_B,\beta_C,\gamma_m,\gamma_p} \rho$$

subject to $P > 0$ and LMIs in (8.23)

$$(8.24)$$

Remark 8.4

i) If the prescribed noise-filtering level ρ is prescribed by a biological engineer, a robust synchronization design would involve specifying the design parameters α_a, α_b, α_c, β_A, β_B, β_C, γ_m, and γ_p of the synthetic gene oscillators in A_k, so that the LMIs in (8.23) have a positive solution $P > 0$ in Proposition 8.2. If we want to achieve optimal filtering of extrinsic noise for the synchronized synthetic oscillators, some design parameters need to be specified for coupled synthetic oscillators to achieve the constrained optimization in (8.24).

ii) In this section, the fuzzy approximation method in (8.21) or (8.22) is only employed to simplify the analysis and design procedure via solving $P > 0$ for LMIs in (8.23) instead of solving $V(e) > 0$ for HJI in (8.16) directly. Further, based on the fuzzy interpolation of local linear systems, i.e., replacing $F(x,s)$, $G(x,s)$, $F_W(x,s)$ and $G_W(x,s)$ by the fuzzy approximations in (8.22), in Proposition 8.2, $V(e) = e^T Pe$ is employed to solve the HJI (8.16) in Proposition 8.1. The HJI in Proposition 8.1 is replaced with a set of LMIs in Proposition 8.2 and we only need to solve $P > 0$ for LMIs to guarantee the coupled synthetic genetic oscillators have a noise filtering level ρ.

iii) In general, the constrained optimization problems in (8.24) are called eigenvalue problem (Boyd et al. 1994), which can be efficiently solved by the MATLAB LMI toolbox.

iv) In addition to the robust oscillation synchronization, the proposed method can be applied to robust synchronization design of coupled synthetic gene networks with any kind of dynamic behavior.

v) In the fuzzy approximation case, the synchronization robustness criterion in (8.19) is equivalent to the following

$$\underbrace{(I_N \otimes A_{Wk} + C \otimes B_{Wk})^T P(I_N \otimes A_{Wk} + C \otimes B_{Wk})}_{\text{local intrinsic robustness}} + R + \underbrace{\frac{1}{\rho_0^2} PHH^T P}_{\text{local extrinsic robustness}} \quad (8.25)$$

$$\leq \underbrace{-\left(P(I_N \otimes A_k + C \otimes B_k) + (I_N \otimes A_k + C \otimes B_k)^T P\right)}_{\text{local synchronization robustness}}$$

for $k = 1, 2, ..., L$ which is equivalent to (8.23) with ρ being replace by ρ_0. The biological meaning of synchronization robustness criterion in (8.25) is that if the local synchronization robustness of local coupled synthetic genetic oscillators can confer local intrinsic robustness to tolerate local intrinsic parameter fluctuation and local extrinsic robustness to filter external noise, then the coupled synthetic genetic oscillators can be synchronized with a noise filtering ability ρ_0. If the synchronization robustness criterion in (8.25) is violated, then the synchronization of coupled synthetic genetic oscillators may not be achieved due to intrinsic parameter fluctuation and extrinsic noise. In general, if the design parameters of coupled synthetic genetic oscillators are specified so that the eigenvalues of local coupled system matrix $I_N \otimes A_{Wk} + C \otimes B_{Wk}$ are far in the left hand side of complex s-domain (i.e., with more negative real part), then the coupled synthetic genetic networks are more easy to synchronize in spite of intrinsic parameter fluctuation and extrinsic noise.

8.5 Robust Synchronization of Synthetic Genetic Oscillators by External Control input

If robust synchronizations of coupled synthetic genetic oscillators cannot be achieved spontaneously via the parameter design in the above sections, then a control strategy is needed from external stimulation inputs to improve the robust synchronization of coupled synthetic genetic oscillators. External stimulation inputs are known to play an important role in the synchronization of biological rhythms. Recently, several methods of periodic stimulation for synchronization of nonlinear oscillators have been introduced (Wang and Chen 2005, Wang et al. 2006). However, even simple methods may show enormous complexity in the control scheme for synchronization of nonlinear stochastic coupled oscillators. In this section, based on nonlinear H_∞ stochastic control theory, an input control strategy is introduced to enhance the robust synchronization. If AI is injected into a common medium to increase the average concentration of AI protein in the extracellular environment, which in turn increases the cellular communication of coupled oscillation systems, then the dynamics of the signaling molecule AI in the cellular environment, as shown in (8.3), should be modified as (Chen and Hsu 2012)

$$\frac{dx_{Si}(t)}{dt} = -d_{se}x_{Si}(t) + \beta_s x_{Ai}(t) - \eta_s \left(Q_e + u_e\right)\left(x_{Si}(t) - N^{-1}\sum_{j=1}^{N} x_{Sj}(t)\right) \qquad (8.26)$$

where $u_e = Q$ represents an extracellular control input, which can be implemented via the injection of inducer AI.

For the simplicity of control design, suppose that the following control input $u_e = Q$ is employed to improve the robust synchronization of the nonlinear stochastic coupled synthetic oscillation systems. In this situation, the synchronization error dynamics in (8.13) should be modified as follows:

$$de = \left(F(x,s) + \left(C_e(Q) \otimes I_m\right)G(x,s) + Hv\right)dt + \left(F_W(x,s) + \left(C_e(Q) \otimes I_m\right)G_W(x,s)\right)dw \quad (8.27)$$

where $C_e(Q) = \left(c_{eij}(Q)\right)_{N \times N} \in R^{N \times N}$ is the coupling configuration matrix, in which $c_{eii}(Q) = -\eta_s\left(1 - N^{-1}\right)\left(Q_e + Q\right)$ if $i = j$, otherwise $c_{eij}(Q) = \eta_s N^{-1}\left(Q_e + Q\right)$. Then, we can also obtain the robust synchronization control design of coupled oscillation systems under intrinsic kinetic parameter fluctuations and extrinsic environmental molecular noises as follows.

Corollary 8.1: For the nonlinear stochastic coupled synthetic genetic oscillators with an extracellular control input $u_e = Q$ in the terms of $C_e(Q)$ in (8.27), if there exists a positive solution $V(e) > 0$ with $V(0) = 0$ to the following HJI

$$e^T Re + \left(\frac{\partial V(e)}{\partial e}\right)^T \left(F(x,s) + \left(C_e(Q) \otimes I_m\right)G(x,s)\right) + \frac{1}{4\rho^2}\left(\frac{\partial V(e)}{\partial e}\right)^T HH^T \left(\frac{\partial V(e)}{\partial e}\right)$$

$$+ \frac{1}{2}\left(F_W(x,s) + \left(C_e(Q) \otimes I_m\right)G_W(x,s)\right)\frac{\partial^2 V(e)}{\partial e^2}\left(F_W(x,s) + \left(C_e(Q) \otimes I_m\right)G_W(x,s)\right) < 0 \quad (8.28)$$

for a prescribed filtering level ρ, then the stochastic intrinsic kinetic noise can be robustly tolerated, and the influence of extrinsic environmental molecular noise $v(t)$ on the synchronization of the nonlinear stochastic coupled synthetic oscillation systems in (8.27) is less than or equal to ρ, i.e., the inequality in (8.14) or (8.15) holds.

Proof: Similar to the proof of Proposition 8.1.

The inequality (8.28) is equivalent to synchronization robustness criterion

$$\underbrace{\frac{1}{2}\left(F_W(x,s) + \left(C_e(Q) \otimes I_m\right)G_W(x,s)\right)\frac{\partial^2 V(e)}{\partial e^2}\left(F_W(x,s) + \left(C_e(Q) \otimes I_m\right)G_W(x,s)\right)}_{\text{intrinsic robustness}}$$

$$\underbrace{+ e^T Re + \frac{1}{4\rho^2}\left(\frac{\partial V(e)}{\partial e}\right)^T HH^T \left(\frac{\partial V(e)}{\partial e}\right)}_{\text{extrinsic robustness}} \leq \underbrace{-\left(\frac{\partial V(e)}{\partial e}\right)^T \left(F(x,s) + \left(C_e(Q) \otimes I_m\right)G(x,s)\right)}_{\text{synchronization robustness}} \quad (8.29)$$

The physical meaning of synchronization robustness criterion in (8.29) is that if we can specify control parameter Q to improve the synchronization robustness to provide more intrinsic robustness and more extrinsic robustness to tolerate more intrinsic parameter fluctuation and filter more extrinsic noise, then the robust synchronization of the nonlinear stochastic coupled synthetic genetic oscillators in (8.27) can be guaranteed. Similarly, the optimal noise-filtering design of synchronized oscillation systems by the extracellular control input in (8.27) can be achieved by solving the following constrained optimization problem:

$$\rho_0 = \min_{Q} \rho \tag{8.30}$$

subject to $V(e) > 0$ and HJI in (8.28)

In general, it is still very difficult to specify the control parameter $u_e = Q$ to solve the HJI-constrained optimization in (8.30) for achieving the optimal noise filtering for synchronized synthetic genetic oscillators. Therefore, the fuzzy approximation method is again employed to simplify the control design procedure. Based on the fuzzy approximation method, the following fuzzy interpolation system is employed to approach the nonlinear stochastic coupled oscillation systems in (8.27):

$$de = \sum_{k=1}^{L} \mu_k(z) \left(\left((I_N \otimes A_k + C_e(Q) \otimes B_k) e + Hv \right) dt + \left((I_N \otimes A_{Wk} + C_e(Q) \otimes B_{Wk}) e \right) dw \right) \tag{8.31}$$

Applying the fuzzy approximation method, the external signal control design can be obtained as described in the following corollary, for robust filtering of synchronized oscillation systems with intrinsic kinetic parameter fluctuations and extrinsic environmental molecular noise.

Corollary 8.2: For stochastic synchronized oscillation systems, if there exists a symmetric solution $P > 0$ to the following LMIs for a prescribed noise-filtering level ρ

$$\begin{bmatrix} R + P(I_N \otimes A_k + C_e(Q) \otimes B_k) + (I_N \otimes A_k + C_e(Q) \otimes B_k)^T P & \\ + (I_N \otimes A_{Wk} + C_e(Q) \otimes B_{Wk})^T P(I_N \otimes A_{Wk} + C_e(Q) \otimes B_{Wk}) & PH \\ H^T P & -\rho^2 I \end{bmatrix} < 0 \tag{8.32}$$

for $k = 1, 2, \cdots, L$, then intrinsic parametric noise can be tolerated and the effect of extrinsic molecular noise $v(t)$ on the synchronization of nonlinear stochastic coupled oscillation systems is less than or equal to a prescribed filtering level ρ.

Proof: Similar to the proof of Proposition 8.2.

The physical meaning of Corollary 8.2 is that if we can select a control parameter Q, such that the LMIs in (8.32) have a positive definite solution $P > 0$, then the robust synchronization with a prescribed noise filtering level ρ on extrinsic environmental molecular noise is guaranteed for the nonlinear stochastic coupled synthetic genetic oscillators. If we specify control parameter Q so that the eigenvalues of local system matrix of

$I_N \otimes A_k + C_e(Q) \otimes B_k$ of coupled synthetic gene oscillators have more negative real part (i.e., in far left hand complex s-domain), the coupled synthetic gene oscillators are with more robust synchronization to tolerate more intrinsic parameter fluctuations and to filter more extrinsic noise.

Similarly, based on the fuzzy approximation method, an optimal noise-filtering design of synchronized oscillation systems by using the extracellular control input in (8.30) can be achieved by solving the following constrained optimization problem:

$$\rho_0 = \min_{Q} \rho$$

subject to $V(e) > 0$ and LMIs in (8.32)   (8.33)

The physical meaning of the constrained optimization in (8.33) is that if we can select a control parameter Q through the inducer concentration control method to solve the constrained optimization problem, we can achieve both robust synchronization against intrinsic kinetic parameter fluctuations and optimal filtering against external environmental molecular noise on the synchronization by using the external control signal in the coupled synthetic oscillation systems.

The design procedure of external inducer control for robust synchronization of the coupled network is summarized as follows (Chen and Hsu 2012):

1) Consider a synthetic genetic network of N coupled oscillators with intrinsic kinetic parameter fluctuations and extrinsic environmental molecular noise.
2) Given the prescribed disturbance attenuation level ρ.
3) Represent the nonlinear stochastic synchronization error dynamic by the T-S fuzzy synchronization error dynamic model, using the interpolation of several local linear stochastic systems.
4) Specify Q to solve LMI in (8.32) with the help of LMI toolbox in MATLAB so that N coupled synthetic genetic oscillators can be synchronized with a prescribed noise filtering level ρ.

8.6 An *in silico* Design Example for Robust Synchronization Design in the Genetic Oscillation Systems

In this section, we provide a simulated example to illustrate the design procedure of robust synchronization of the nonlinear stochastic coupled synthetic oscillation systems and to confirm the performance of the robust synchronization of proposed method against intrinsic kinetic parameter fluctuations and extrinsic environmental molecular noise.

The purpose of this example is to demonstrate the effectiveness of the theoretical synchronization result of synthetic gene oscillators in mimicking real biological oscillator systems. We consider a synthetic genetic network of $N = 10$ coupled synthetic genetic oscillators with intrinsic kinetic parameter fluctuations and extrinsic environmental molecular noise in (8.5). The simulation results are shown in Figure 8.2. It can be seen that the parameter set of the coupled synthetic oscillator network, as listed in the figure legend, cannot make the whole network synchronize spontaneously. Suppose we want to specify a control parameter Q (which is proportional to the density of inducer AI) in (8.26) to compensate for the inefficiency of coupling between the synthetic genetic oscillators from the quasi-steady-state point of view. The design procedure first begins with representing the nonlinear stochastic synchronization error dynamic in (8.22) by the T-S fuzzy

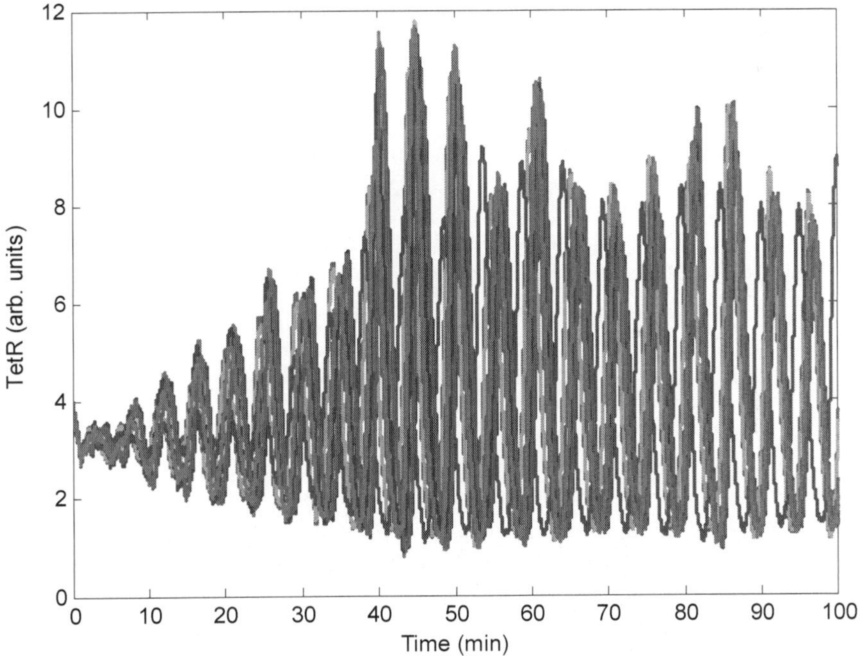

Figure 8.2. Ten coupled genetic oscillators. The parameter values in (8.1), (8.2), and (8.3) are set as follows: $\alpha_a=\alpha_b=\alpha_c=216$, $u_s=20$, $\mu-1.2$, μ_s-1, $n-2$, $\gamma_s=1$, η_s-2, $\beta_s=0.1$, $\beta_A-\beta_B-\beta_C-1$, $\gamma_m=6.9315$, $\gamma_p=1.1552$ and $Q_e=0.09$. Suppose the nonlinear stochastic coupled synthetic oscillators suffer from stochastic parameter fluctuations as shown in (8.5) with $\Delta\alpha_a=\Delta\alpha_b=\Delta\alpha_c=2.16$, $\Delta\alpha_s=0.2$, $\Delta\beta_A=\Delta\beta_B=\Delta\beta_C=0.01$, $\Delta\beta_s=0.001$, $\Delta\eta_s=0.02$, $\Delta\gamma_m=0.06$, $\Delta\gamma_p=0.01$, and $\Delta\gamma_s=0.01$. For the convenience of simulation, we assume that the extrinsic molecular noise $v_1{\sim}v_{10}$ is independent Gaussian white noise with a mean of zero and standard deviation of 0.02. It can be seen that coupled synthetic oscillators cannot achieve synchronization under these intrinsic kinetic parameter fluctuations and extrinsic molecular noise.

Color image of this figure appears in the color plate section at the end of the book.

synchronization error dynamic model in (8.31), using the interpolation of several linear stochastic systems as presented in Appendix 8.3. According to the fuzzy approximation and Corollary 8.2, our control design problem is how to specify Q (i.e., the corresponding density of inducer AI), so that the ten coupled synthetic genetic oscillators have a positive solution $P > 0$ with a prescribed noise filtering level $\rho = 0.56$ to guarantee robust synchronization under intrinsic kinetic parameter fluctuations and extrinsic environmental molecular noise. The LMI toolbox in MATLAB can then be used to significantly simplify the system analysis and design procedure. With $Q = 0.66$ solved from LMIs in (8.32), the outputs of the coupled gene network of ten synthetic oscillators under intrinsic parametric fluctuations and extrinsic noise are shown in Figure 8.3. It can be seen that the coupled synthetic genetic oscillators have robust synchronizability to achieve the synchronous behavior despite the effect of uncertain initial state, intrinsic kinetic parameter fluctuations, and extrinsic environmental molecular noise on the host cell. According to a Monte Carlo simulation with 100 runs, the noise-filtering level of the coupled gene network is given by

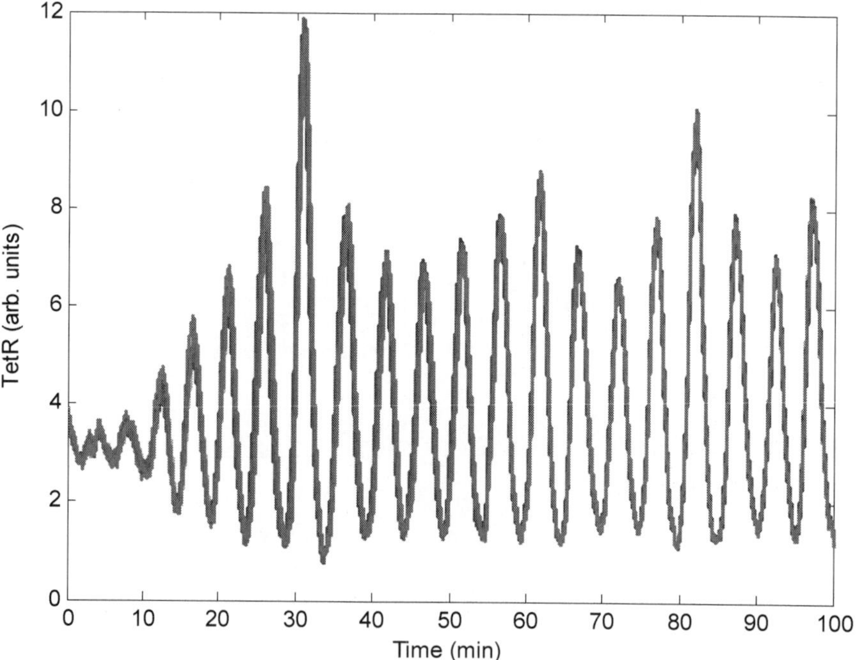

Figure 8.3. Robust synchronization result of ten coupled synthetic oscillators in Figure 8.2, by external control with $Q = 0.66$.

Color image of this figure appears in the color plate section at the end of the book.

$\left(E \int_{0}^{100} e^{T}(t) Re(t) dt \right) \Big/ \left(E \int_{0}^{100} v^{T}(t) v(t) dt \right) \approx 0.19^{2} < 0.56^{2}$. It can be clearly seen that based on our proposed design method, the coupled gene network can not only tolerate kinetic parameter variations but also attenuate the extrinsic molecular noise below a desired level to achieve a robust synchronization.

8.7 Summary

The cell is the functional unit of all living things, either unicellular or multicellular. A cell can sense many different signals from the internal or external context and can respond to the constantly changing environment via appropriate cellular processes. Also, cells can interact with each other via cell-to-cell communication and achieve specific physiological functions essential for life in a cooperative manner. However, many fundamental questions remain regarding how cellular phenomena arise from the interactions between genes and proteins, what features make the cell operate reliably in diverse conditions, and how the cell is responsible for these operations. To gain insight into these questions, one can construct the underlying mechanisms that constitute the web of interactions. This idea is useful to separate a complicated network into many simpler ones, which can work independently but also cooperate with each other. It may not only enhance our understanding of collective behavior particularly via synchronization but may also establish a foundation for robust implementation of coupled synthetic gene networks.

In this chapter, we consider a nonlinear stochastic coupled network with two or more coupled synthetic oscillators. By transforming nonlinear stochastic coupled network dynamics into synchronization error dynamics, we can use Lyapunov's direct method to infer a sufficient condition required for robustness of the nonlinear synchronized network. Assuming that each synthetic oscillator suffers from intrinsic kinetic parameter fluctuations and extrinsic molecular noise, robust synchronization performance is defined as the effect of extrinsic molecular noise upon the synchronization error. Based on this definition, robust synchronization performance of a nonlinear coupled network can be calculated by solving an associated HJI-constrained optimization problem. We also show that nonlinear coupled networks with robust synchronization performance are also synchronizable. Based on this synchronization performance, we propose a procedure for designing or compensating a coupled network with two or more coupled synthetic oscillator through a given connected topology toward a desired robust synchronization performance. Using the proposed method, the coupled synthetic oscillators can not only tolerate kinetic parameter variations but also attenuate the extrinsic molecular noise below a desired level to achieve a desired robust synchronization. However, the HJI-constrained optimization problem is difficult to solve directly by any analytical or numerical method because of the complexity of nonlinear synchronization error dynamics. Hence, we employ a T–S fuzzy model to solve the HJI easily and indirectly. The T–S fuzzy model has been widely applied to approximate nonlinear systems by interpolating several local linearized systems. Here, we use the T–S fuzzy model to approximate the nonlinear stochastic synchronization error dynamics. By using the T–S fuzzy model and choosing the appropriate Lyapunov function, the HJI-constrained optimization for calculating the robust synchronization performance of a nonlinear coupled network is reduced to an equivalent LMI-constrained optimization problem, which can be solved efficiently by MATLAB's LMI toolbox. In such situations, our proposed evaluation framework may offer a possible guideline for the design or compensation of such coupled networks with a given connected topology toward a desired collective behavior.

Appendix 8.1: Proof of Proposition 8.1

Consider the following equality

$$E \int_0^{t_f} e^T R e\, dt = EV\left(e(0)\right) - EV\left(e(t_f)\right) + E \int_0^{t_f} \left(e^T R e + \frac{dV(e)}{dt} \right) dt \qquad (8.34)$$

By Ito formula, we get

$$dV(e)$$

$$= \left(\frac{\partial V(e)}{\partial e} \right)^T \left(\left(F(x,s) + (C \otimes I_m) G(x,s) + Hv \right) dt + \left(F_W(x,s) + (C \otimes I_m) G_W(x,s) \right) dw \right) \qquad (8.35)$$

$$+ \frac{1}{2} \left(F_W(x,s) + (C \otimes I_m) G_W(x,s) \right)^T \frac{\partial^2 V(e)}{\partial e^2} \left(F_W(x,s) + (C \otimes I_m) G_W(x,s) \right) dt$$

Substituting (8.35) into (8.34), and by the fact $EdW = 0, EV\left(e(t_f)\right) \geq 0$, we get

$$E \int_0^{t_f} e^T R e\, dt \leq EV\left(e(0)\right) + E \int_0^{t_f} \left(e^T R e + \left(\frac{\partial V(e)}{\partial e} \right)^T \left(F(x,s) + (C \otimes I_m) G(x,s) + Hv \right) \right.$$

$$\left. + \frac{1}{2} \left(F_W(x,s) + (C \otimes I_m) G_W(x,s) \right)^T \frac{\partial^2 V(e)}{\partial e^2} \left(F_W(x,s) + (C \otimes I_m) G_W(x,s) \right) \right) dt \qquad (8.36)$$

By the fact that

$$\mathbf{a}^T \mathbf{b} = 2 \left(\frac{1}{2\rho} \mathbf{a} \right)^T (\rho \mathbf{b}) \leq \frac{1}{4\rho^2} \mathbf{a}^T \mathbf{a} + \rho^2 \mathbf{b}^T \mathbf{b} \qquad (8.37)$$

for any $\rho > 0$ and vectors $\mathbf{a}$ and $\mathbf{b}$. Then, we get the following inequality

$$E \int_0^{t_f} e^T R e \, dt \le E V\left(e(0)\right)$$

$$+ E \int_0^{t_f} \left[\left(e^T R e + \left(\frac{\partial V(e)}{\partial e} \right)^T \right) \left(F(x,s) + (C \otimes I_m) G(x,s) \right) + \frac{1}{4\rho^2} \left(\frac{\partial V(e)}{\partial e} \right)^T H H^T \left(\frac{\partial V(e)}{\partial e} \right) \right.$$

$$\left. + \frac{1}{2} \left(F_W(x,s) + (C \otimes I_m) G_W(x,s) \right) \frac{\partial^2 V(e)}{\partial e^2} \left(F_W(x,s) + (C \otimes I_m) G_W(x,s) \right) \right] dt$$

where $\mathbf{a} = H^T \dfrac{\partial V(e)}{\partial e}$ and $\mathbf{b} = v$

By the inequality in (8.19), we get

$$E \int_0^{t_f} e^T R e \, dt \le E V\left(e(0)\right) + E \rho^2 \int_0^{t_f} v^T v \, dt \tag{8.38}$$

If coupled synthetic genetic network is free of extrinsic noise, i.e., $v(t) = 0$, then from (8.38), we get

$$E \int_0^{t_f} e^T R e \, dt \le E V\left(e(0)\right) \tag{8.39}$$

Since $EV(e(0))$ is a constant, from (8.39), we get $E(e(t)) \to 0$ as $t_f \to \infty$ and intrinsic parameter fluctuation is tolerated.

Appendix 8.2: Proof of Proposition 8.2

First, we use the following fuzzy interpolation system

$$de = \sum_{k=1}^{N} \mu_k(z) \left(\left((I_N \otimes A_k + C \otimes B_k) e + Hv \right) dt + \left((I_N \otimes A_{Wk} + C \otimes B_{Wk}) e \right) dw \right)$$

in (8.24) to replace (8.16). Following the proof of Proposition 8.1 in Appendix 8.1, we get the following result

$$E \int_0^{t_f} e^T R e \, dt \le E V\left(e(0)\right) - EV\left(e(t_f)\right)$$

$$+ E \int_0^{t_f} \left[e^T R e + \sum_{k=1}^{L} \mu_k(z) \left(\left(\frac{\partial V(e)}{\partial e} \right)^T \left((I_N \otimes A_k + C \otimes B_k) e + Hv \right) \right. \right.$$

$$\left. \left. + \frac{1}{2} e^T (I_N \otimes A_{Wk} + C \otimes B_{Wk}) \frac{\partial^2 V(e)}{\partial e^2} (I_N \otimes A_{Wk} + C \otimes B_{Wk}) e \right) \right] dt \tag{8.40}$$

By the facts that $V\left(e(t_f)\right) \ge 0$, and

$$\left(\frac{\partial V(e)}{\partial e} \right)^T Hv \le \frac{1}{4\rho^2} \left(\frac{\partial V(e)}{\partial e} \right)^T H H^T \left(\frac{\partial V(e)}{\partial e} \right) + \rho^2 v^T v$$

we get the following result from (8.40)

$$E \int_0^{t_f} e^T Re\, dt \leq EV(e(0))$$

$$+E \int_0^{t_f} \left(e^T Re + \sum_{k=1}^{L} \mu_k(z) \left(\left(\frac{\partial V(e)}{\partial e} \right)^T (I_N \otimes A_k + C \otimes B_k)e + \frac{1}{4\rho^2} \left(\frac{\partial V(e)}{\partial e} \right)^T HH^T \left(\frac{\partial V(e)}{\partial e} \right) \right) \right.$$

$$\left. + \rho^2 v^T v + \frac{1}{2} e^T (I_N \otimes A_{Wk} + C \otimes B_{Wk})^T \frac{\partial^2 V(e)}{\partial e^2} (I_N \otimes A_{Wk} + C \otimes B_{Wk})e \right) dt \right) \quad (8.41)$$

If we choose $V(e) = e^T Pe$, then $\dfrac{\partial V(e)}{\partial e} = 2Pe$, $\dfrac{\partial^2 V(e)}{\partial e^2} = 2P$, then we get the following result from (8.41)

$$E \int_0^{t_f} e^T Re\, dt \leq Ee(0)^T Pe(0)$$

$$+ E \sum_{k=1}^{L} \mu_k(z) \int_0^{t_f} e^T \left(R + P(I_N \otimes A_k + C \otimes B_k) + (I_N \otimes A_k + C \otimes B_k)^T P \right.$$

$$\left. + (I_N \otimes A_{Wk} + C \otimes B_{Wk})^T P(I_N \otimes A_{Wk} + C \otimes B_{Wk}) + \frac{1}{\rho^2} PHH^T P \right) e\, dt + \rho^2 v^T v\, dt \quad (8.42)$$

If

$$R + P(I_N \otimes A_k + C \otimes B_k) + (I_N \otimes A_k + C \otimes B_k)^T P$$

$$+ (I_N \otimes A_{Wk} + C \otimes B_{Wk})^T P(I_N \otimes A_{Wk} + C \otimes B_{Wk}) + \frac{1}{\rho^2} PHH^T P < 0 \quad (8.43)$$

then we get

$$E \int_0^{t_f} e^T Re\, dt \leq Ee(0)^T Pe(0) + E\rho^2 \int_0^{t_f} v^T v\, dt$$

which is (8.18) and will be reduced to (8.17) if $e(0) = 0$.

Therefore, if the LMIs in (8.43) hold, then the noise filtering ability ρ on the synchronization of the coupled oscillation systems is achieved. By Schur complement, the inequalities in (8.43) are equivalent to the LMIs in (8.25), i.e., if the LMIs in (8.25) have a common solution $P > 0$, then the synchronization of nonlinear stochastic coupled synthetic oscillation systems in (8.8) has a filtering level ρ against the extrinsic noises.

Appendix 8.3

The fuzzy approximation is employed to approximate the nonlinear stochastic synchronization error dynamic by interpolating several local linear stochastic systems as follows:

$$de = \left(F(x,s) + (C \otimes I_m)G(x,s) + Hv\right)dt + \left(F_W(x,s) + (C \otimes I_m)G_W(x,s)\right)dw$$

$$= \sum_{k=1}^{16} \mu_k(z)\left(\left((I_N \otimes A_k + C \otimes B)e + Hv\right)dt + \left((I_N \otimes A_{Wk} + C \otimes B_W)e\right)dw\right)$$

where

$$A_1 = \begin{bmatrix}
-6.93 & 0 & 0 & 0 & 0 & -86.4 & 0 \\
0 & -6.93 & 0 & -86.4 & 0 & 0 & 0 \\
0 & 0 & -6.93 & 0 & -0.15 & 0 & 9.90 \\
1.00 & 0 & 0 & -1.15 & 0 & 0 & 0 \\
0 & 1.00 & 0 & 0 & -1.15 & 0 & 0 \\
0 & 0 & 1.00 & 0 & 0 & -1.15 & 0 \\
0 & 0 & 0 & 0.10 & 0 & 0 & -1.00
\end{bmatrix}$$

$$A_{W1} = 0.01 * \begin{bmatrix}
-6.93 & 0 & 0 & 0 & 0 & -86.4 & 0 \\
0 & -6.93 & 0 & -86.4 & 0 & 0 & 0 \\
0 & 0 & -6.93 & 0 & -0.15 & 0 & 9.90 \\
1.00 & 0 & 0 & -1.15 & 0 & 0 & 0 \\
0 & 1.00 & 0 & 0 & -1.15 & 0 & 0 \\
0 & 0 & 1.00 & 0 & 0 & -1.15 & 0 \\
0 & 0 & 0 & 0.10 & 0 & 0 & -1.00
\end{bmatrix}$$

$$A_2 = \begin{bmatrix}
-6.93 & 0 & 0 & 0 & 0 & -86.4 & 0 \\
0 & -6.93 & 0 & -86.4 & 0 & 0 & 0 \\
0 & 0 & -6.93 & 0 & -0.15 & 0 & 8.65 \\
1.00 & 0 & 0 & -1.15 & 0 & 0 & 0 \\
0 & 1.00 & 0 & 0 & -1.15 & 0 & 0 \\
0 & 0 & 1.00 & 0 & 0 & -1.15 & 0 \\
0 & 0 & 0 & 0.10 & 0 & 0 & -1.00
\end{bmatrix}$$

$$A_{W2} = 0.01 * \begin{bmatrix}
-6.93 & 0 & 0 & 0 & 0 & -86.4 & 0 \\
0 & -6.93 & 0 & -86.4 & 0 & 0 & 0 \\
0 & 0 & -6.93 & 0 & -0.15 & 0 & 8.65 \\
1.00 & 0 & 0 & -1.15 & 0 & 0 & 0 \\
0 & 1.00 & 0 & 0 & -1.15 & 0 & 0 \\
0 & 0 & 1.00 & 0 & 0 & -1.15 & 0 \\
0 & 0 & 0 & 0.10 & 0 & 0 & -1.00
\end{bmatrix}$$

$$A_3 = \begin{bmatrix} -6.93 & 0 & 0 & 0 & 0 & -86.4 & 0 \\ 0 & -6.93 & 0 & -86.4 & 0 & 0 & 0 \\ 0 & 0 & -6.93 & 0 & -0.00 & 0 & 9.90 \\ 1.00 & 0 & 0 & -1.15 & 0 & 0 & 0 \\ 0 & 1.00 & 0 & 0 & -1.15 & 0 & 0 \\ 0 & 0 & 1.00 & 0 & 0 & -1.15 & 0 \\ 0 & 0 & 0 & 0.10 & 0 & 0 & -1.00 \end{bmatrix}$$

$$A_{W3} = 0.01 * \begin{bmatrix} -6.93 & 0 & 0 & 0 & 0 & -86.4 & 0 \\ 0 & -6.93 & 0 & -86.4 & 0 & 0 & 0 \\ 0 & 0 & -6.93 & 0 & -0.00 & 0 & 9.90 \\ 1.00 & 0 & 0 & -1.15 & 0 & 0 & 0 \\ 0 & 1.00 & 0 & 0 & -1.15 & 0 & 0 \\ 0 & 0 & 1.00 & 0 & 0 & -1.15 & 0 \\ 0 & 0 & 0 & 0.10 & 0 & 0 & -1.00 \end{bmatrix}$$

$$A_4 = \begin{bmatrix} -6.93 & 0 & 0 & 0 & 0 & -86.4 & 0 \\ 0 & -6.93 & 0 & -86.4 & 0 & 0 & 0 \\ 0 & 0 & -6.93 & 0 & -0.00 & 0 & 8.65 \\ 1.00 & 0 & 0 & -1.15 & 0 & 0 & 0 \\ 0 & 1.00 & 0 & 0 & -1.15 & 0 & 0 \\ 0 & 0 & 1.00 & 0 & 0 & -1.15 & 0 \\ 0 & 0 & 0 & 0.10 & 0 & 0 & -1.00 \end{bmatrix}$$

$$A_{W4} = 0.01 * \begin{bmatrix} -6.93 & 0 & 0 & 0 & 0 & -86.4 & 0 \\ 0 & -6.93 & 0 & -86.4 & 0 & 0 & 0 \\ 0 & 0 & -6.93 & 0 & -0.00 & 0 & 8.65 \\ 1.00 & 0 & 0 & -1.15 & 0 & 0 & 0 \\ 0 & 1.00 & 0 & 0 & -1.15 & 0 & 0 \\ 0 & 0 & 1.00 & 0 & 0 & -1.15 & 0 \\ 0 & 0 & 0 & 0.10 & 0 & 0 & -1.00 \end{bmatrix}$$

$$A_5 = \begin{bmatrix} -6.93 & 0 & 0 & 0 & 0 & -86.4 & 0 \\ 0 & -6.93 & 0 & -0.001 & 0 & 0 & 0 \\ 0 & 0 & -6.93 & 0 & -0.15 & 0 & 9.90 \\ 1.00 & 0 & 0 & -1.15 & 0 & 0 & 0 \\ 0 & 1.00 & 0 & 0 & -1.15 & 0 & 0 \\ 0 & 0 & 1.00 & 0 & 0 & -1.15 & 0 \\ 0 & 0 & 0 & 0.10 & 0 & 0 & -1.00 \end{bmatrix}$$

$$A_{W5} = 0.01 * \begin{bmatrix} -6.93 & 0 & 0 & 0 & 0 & -86.4 & 0 \\ 0 & -6.93 & 0 & -0.001 & 0 & 0 & 0 \\ 0 & 0 & -6.93 & 0 & -0.15 & 0 & 9.90 \\ 1.00 & 0 & 0 & -1.15 & 0 & 0 & 0 \\ 0 & 1.00 & 0 & 0 & -1.15 & 0 & 0 \\ 0 & 0 & 1.00 & 0 & 0 & -1.15 & 0 \\ 0 & 0 & 0 & 0.10 & 0 & 0 & -1.00 \end{bmatrix}$$

$$A_6 = \begin{bmatrix} -6.93 & 0 & 0 & 0 & 0 & -86.4 & 0 \\ 0 & -6.93 & 0 & -0.001 & 0 & 0 & 0 \\ 0 & 0 & -6.93 & 0 & -0.15 & 0 & 8.65 \\ 1.00 & 0 & 0 & -1.15 & 0 & 0 & 0 \\ 0 & 1.00 & 0 & 0 & -1.15 & 0 & 0 \\ 0 & 0 & 1.00 & 0 & 0 & -1.15 & 0 \\ 0 & 0 & 0 & 0.10 & 0 & 0 & -1.00 \end{bmatrix}$$

$$A_{W6} = 0.01 * \begin{bmatrix} -6.93 & 0 & 0 & 0 & 0 & -86.4 & 0 \\ 0 & -6.93 & 0 & -0.001 & 0 & 0 & 0 \\ 0 & 0 & -6.93 & 0 & -0.15 & 0 & 9.90 \\ 1.00 & 0 & 0 & -1.15 & 0 & 0 & 0 \\ 0 & 1.00 & 0 & 0 & -1.15 & 0 & 0 \\ 0 & 0 & 1.00 & 0 & 0 & -1.15 & 0 \\ 0 & 0 & 0 & 0.10 & 0 & 0 & -1.00 \end{bmatrix}$$

$$A_7 = \begin{bmatrix} -6.93 & 0 & 0 & 0 & 0 & -86.4 & 0 \\ 0 & -6.93 & 0 & -0.001 & 0 & 0 & 0 \\ 0 & 0 & -6.93 & 0 & -0.00 & 0 & 9.90 \\ 1.00 & 0 & 0 & -1.15 & 0 & 0 & 0 \\ 0 & 1.00 & 0 & 0 & -1.15 & 0 & 0 \\ 0 & 0 & 1.00 & 0 & 0 & -1.15 & 0 \\ 0 & 0 & 0 & 0.10 & 0 & 0 & -1.00 \end{bmatrix}$$

$$A_{W7} = 0.01 * \begin{bmatrix} -6.93 & 0 & 0 & 0 & 0 & -86.4 & 0 \\ 0 & -6.93 & 0 & -0.001 & 0 & 0 & 0 \\ 0 & 0 & -6.93 & 0 & -0.00 & 0 & 9.90 \\ 1.00 & 0 & 0 & -1.15 & 0 & 0 & 0 \\ 0 & 1.00 & 0 & 0 & -1.15 & 0 & 0 \\ 0 & 0 & 1.00 & 0 & 0 & -1.15 & 0 \\ 0 & 0 & 0 & 0.10 & 0 & 0 & -1.00 \end{bmatrix}$$

$$
A_8 = \begin{bmatrix}
-6.93 & 0 & 0 & 0 & 0 & -86.4 & 0 \\
0 & -6.93 & 0 & -0.001 & 0 & 0 & 0 \\
0 & 0 & -6.93 & 0 & -0.00 & 0 & 8.65 \\
1.00 & 0 & 0 & -1.15 & 0 & 0 & 0 \\
0 & 1.00 & 0 & 0 & -1.15 & 0 & 0 \\
0 & 0 & 1.00 & 0 & 0 & -1.15 & 0 \\
0 & 0 & 0 & 0.10 & 0 & 0 & -1.00
\end{bmatrix}
$$

$$
A_{W8} = 0.01* \begin{bmatrix}
-6.93 & 0 & 0 & 0 & 0 & -86.4 & 0 \\
0 & -6.93 & 0 & -0.001 & 0 & 0 & 0 \\
0 & 0 & -6.93 & 0 & -0.00 & 0 & 8.65 \\
1.00 & 0 & 0 & -1.15 & 0 & 0 & 0 \\
0 & 1.00 & 0 & 0 & -1.15 & 0 & 0 \\
0 & 0 & 1.00 & 0 & 0 & -1.15 & 0 \\
0 & 0 & 0 & 0.10 & 0 & 0 & -1.00
\end{bmatrix}
$$

$$
A_9 = \begin{bmatrix}
-6.93 & 0 & 0 & 0 & 0 & -0.001 & 0 \\
0 & -6.93 & 0 & -86.4 & 0 & 0 & 0 \\
0 & 0 & -6.93 & 0 & -0.15 & 0 & 9.90 \\
1.00 & 0 & 0 & -1.15 & 0 & 0 & 0 \\
0 & 1.00 & 0 & 0 & -1.15 & 0 & 0 \\
0 & 0 & 1.00 & 0 & 0 & -1.15 & 0 \\
0 & 0 & 0 & 0.10 & 0 & 0 & -1.00
\end{bmatrix}
$$

$$
A_{W9} = 0.01* \begin{bmatrix}
-6.93 & 0 & 0 & 0 & 0 & -0.001 & 0 \\
0 & -6.93 & 0 & -86.4 & 0 & 0 & 0 \\
0 & 0 & -6.93 & 0 & -0.15 & 0 & 9.90 \\
1.00 & 0 & 0 & -1.15 & 0 & 0 & 0 \\
0 & 1.00 & 0 & 0 & -1.15 & 0 & 0 \\
0 & 0 & 1.00 & 0 & 0 & -1.15 & 0 \\
0 & 0 & 0 & 0.10 & 0 & 0 & -1.00
\end{bmatrix}
$$

$$
A_{10} = \begin{bmatrix}
-6.93 & 0 & 0 & 0 & 0 & -0.001 & 0 \\
0 & -6.93 & 0 & -86.4 & 0 & 0 & 0 \\
0 & 0 & -6.93 & 0 & -0.15 & 0 & 8.65 \\
1.00 & 0 & 0 & -1.15 & 0 & 0 & 0 \\
0 & 1.00 & 0 & 0 & -1.15 & 0 & 0 \\
0 & 0 & 1.00 & 0 & 0 & -1.15 & 0 \\
0 & 0 & 0 & 0.10 & 0 & 0 & -1.00
\end{bmatrix}
$$

$$A_{W10} = 0.01 * \begin{bmatrix} -6.93 & 0 & 0 & 0 & 0 & -0.001 & 0 \\ 0 & -6.93 & 0 & -86.4 & 0 & 0 & 0 \\ 0 & 0 & -6.93 & 0 & -0.15 & 0 & 8.65 \\ 1.00 & 0 & 0 & -1.15 & 0 & 0 & 0 \\ 0 & 1.00 & 0 & 0 & -1.15 & 0 & 0 \\ 0 & 0 & 1.00 & 0 & 0 & -1.15 & 0 \\ 0 & 0 & 0 & 0.10 & 0 & 0 & -1.00 \end{bmatrix}$$

$$A_{11} = \begin{bmatrix} -6.93 & 0 & 0 & 0 & 0 & -0.001 & 0 \\ 0 & -6.93 & 0 & -86.4 & 0 & 0 & 0 \\ 0 & 0 & -6.93 & 0 & -0.00 & 0 & 9.90 \\ 1.00 & 0 & 0 & -1.15 & 0 & 0 & 0 \\ 0 & 1.00 & 0 & 0 & -1.15 & 0 & 0 \\ 0 & 0 & 1.00 & 0 & 0 & -1.15 & 0 \\ 0 & 0 & 0 & 0.10 & 0 & 0 & -1.00 \end{bmatrix}$$

$$A_{W11} = 0.01 * \begin{bmatrix} -6.93 & 0 & 0 & 0 & 0 & -0.001 & 0 \\ 0 & -6.93 & 0 & -86.4 & 0 & 0 & 0 \\ 0 & 0 & -6.93 & 0 & -0.00 & 0 & 9.90 \\ 1.00 & 0 & 0 & -1.15 & 0 & 0 & 0 \\ 0 & 1.00 & 0 & 0 & -1.15 & 0 & 0 \\ 0 & 0 & 1.00 & 0 & 0 & -1.15 & 0 \\ 0 & 0 & 0 & 0.10 & 0 & 0 & -1.00 \end{bmatrix}$$

$$A_{12} = \begin{bmatrix} -6.93 & 0 & 0 & 0 & 0 & -0.001 & 0 \\ 0 & -6.93 & 0 & -86.4 & 0 & 0 & 0 \\ 0 & 0 & -6.93 & 0 & -0.00 & 0 & 8.65 \\ 1.00 & 0 & 0 & -1.15 & 0 & 0 & 0 \\ 0 & 1.00 & 0 & 0 & -1.15 & 0 & 0 \\ 0 & 0 & 1.00 & 0 & 0 & -1.15 & 0 \\ 0 & 0 & 0 & 0.10 & 0 & 0 & -1.00 \end{bmatrix}$$

$$A_{W12} = 0.01 * \begin{bmatrix} -6.93 & 0 & 0 & 0 & 0 & -0.001 & 0 \\ 0 & -6.93 & 0 & -86.4 & 0 & 0 & 0 \\ 0 & 0 & -6.93 & 0 & -0.00 & 0 & 8.65 \\ 1.00 & 0 & 0 & -1.15 & 0 & 0 & 0 \\ 0 & 1.00 & 0 & 0 & -1.15 & 0 & 0 \\ 0 & 0 & 1.00 & 0 & 0 & -1.15 & 0 \\ 0 & 0 & 0 & 0.10 & 0 & 0 & -1.00 \end{bmatrix}$$

$$A_{13} = \begin{bmatrix} -6.93 & 0 & 0 & 0 & 0 & -0.001 & 0 \\ 0 & -6.93 & 0 & -0.001 & 0 & 0 & 0 \\ 0 & 0 & -6.93 & 0 & -0.15 & 0 & 9.90 \\ 1.00 & 0 & 0 & -1.15 & 0 & 0 & 0 \\ 0 & 1.00 & 0 & 0 & -1.15 & 0 & 0 \\ 0 & 0 & 1.00 & 0 & 0 & -1.15 & 0 \\ 0 & 0 & 0 & 0.10 & 0 & 0 & -1.00 \end{bmatrix}$$

$$A_{W13} = 0.01 * \begin{bmatrix} -6.93 & 0 & 0 & 0 & 0 & -0.001 & 0 \\ 0 & -6.93 & 0 & -0.001 & 0 & 0 & 0 \\ 0 & 0 & -6.93 & 0 & -0.15 & 0 & 9.90 \\ 1.00 & 0 & 0 & -1.15 & 0 & 0 & 0 \\ 0 & 1.00 & 0 & 0 & -1.15 & 0 & 0 \\ 0 & 0 & 1.00 & 0 & 0 & -1.15 & 0 \\ 0 & 0 & 0 & 0.10 & 0 & 0 & -1.00 \end{bmatrix}$$

$$A_{14} = \begin{bmatrix} -6.93 & 0 & 0 & 0 & 0 & -0.001 & 0 \\ 0 & -6.93 & 0 & -0.001 & 0 & 0 & 0 \\ 0 & 0 & -6.93 & 0 & -0.15 & 0 & 8.65 \\ 1.00 & 0 & 0 & -1.15 & 0 & 0 & 0 \\ 0 & 1.00 & 0 & 0 & -1.15 & 0 & 0 \\ 0 & 0 & 1.00 & 0 & 0 & -1.15 & 0 \\ 0 & 0 & 0 & 0.10 & 0 & 0 & -1.00 \end{bmatrix}$$

$$A_{W14} = 0.01 * \begin{bmatrix} -6.93 & 0 & 0 & 0 & 0 & -0.001 & 0 \\ 0 & -6.93 & 0 & -0.001 & 0 & 0 & 0 \\ 0 & 0 & -6.93 & 0 & -0.15 & 0 & 8.65 \\ 1.00 & 0 & 0 & -1.15 & 0 & 0 & 0 \\ 0 & 1.00 & 0 & 0 & -1.15 & 0 & 0 \\ 0 & 0 & 1.00 & 0 & 0 & -1.15 & 0 \\ 0 & 0 & 0 & 0.10 & 0 & 0 & -1.00 \end{bmatrix}$$

$$A_{15} = \begin{bmatrix} -6.93 & 0 & 0 & 0 & 0 & -0.001 & 0 \\ 0 & -6.93 & 0 & -0.001 & 0 & 0 & 0 \\ 0 & 0 & -6.93 & 0 & -0.00 & 0 & 9.90 \\ 1.00 & 0 & 0 & -1.15 & 0 & 0 & 0 \\ 0 & 1.00 & 0 & 0 & -1.15 & 0 & 0 \\ 0 & 0 & 1.00 & 0 & 0 & -1.15 & 0 \\ 0 & 0 & 0 & 0.10 & 0 & 0 & -1.00 \end{bmatrix}$$

$$A_{W15} = 0.01 * \begin{bmatrix} -6.93 & 0 & 0 & 0 & 0 & -0.001 & 0 \\ 0 & -6.93 & 0 & -0.001 & 0 & 0 & 0 \\ 0 & 0 & -6.93 & 0 & -0.00 & 0 & 9.90 \\ 1.00 & 0 & 0 & -1.15 & 0 & 0 & 0 \\ 0 & 1.00 & 0 & 0 & -1.15 & 0 & 0 \\ 0 & 0 & 1.00 & 0 & 0 & -1.15 & 0 \\ 0 & 0 & 0 & 0.10 & 0 & 0 & -1.00 \end{bmatrix}$$

$$A_{16} = \begin{bmatrix} -6.93 & 0 & 0 & 0 & 0 & -0.001 & 0 \\ 0 & -6.93 & 0 & -0.001 & 0 & 0 & 0 \\ 0 & 0 & -6.93 & 0 & -0.00 & 0 & 8.65 \\ 1.00 & 0 & 0 & -1.15 & 0 & 0 & 0 \\ 0 & 1.00 & 0 & 0 & -1.15 & 0 & 0 \\ 0 & 0 & 1.00 & 0 & 0 & -1.15 & 0 \\ 0 & 0 & 0 & 0.10 & 0 & 0 & -1.00 \end{bmatrix}$$

$$A_{W16} = 0.01 * \begin{bmatrix} -6.93 & 0 & 0 & 0 & 0 & -0.001 & 0 \\ 0 & -6.93 & 0 & -0.001 & 0 & 0 & 0 \\ 0 & 0 & -6.93 & 0 & -0.00 & 0 & 8.65 \\ 1.00 & 0 & 0 & -1.15 & 0 & 0 & 0 \\ 0 & 1.00 & 0 & 0 & -1.15 & 0 & 0 \\ 0 & 0 & 1.00 & 0 & 0 & -1.15 & 0 \\ 0 & 0 & 0 & 0.10 & 0 & 0 & -1.00 \end{bmatrix}$$

References

Boyd, S., El Ghaoui, L., Feron, E. and Balakrishnan, V. 1994. Linear Matrix Inequalities in System and Control Theory. Society for Industrial Mathematics, Philadelphia.

Chen, B.S., Tseng, C.S. and Uang, H.J. 1999. Robustness design of nonlinear dynamic systems via fuzzy linear control. IEEE Transactions on Fuzzy Systems 7: 571–585.

Chen, B.S. and Zhang, W. 2004. Stochastic H-2/H-infinity control with state-dependent noise. IEEE Transactions on Automatic Control 49: 45–57.

Chen, B.S. and Wang, Y.C. 2006. On the attenuation and amplification of molecular noise in genetic regulatory networks. BMC Bioinformatics 7: 52.

Chen, B.S. and Wu, W.S. 2008. Robust filtering circuit design for stochastic gene networks under intrinsic and extrinsic molecular noises. Math Biosci 211: 342–355.

Chen, B.S. and Hsu, C.Y. 2012. Robust synchronization control scheme of a population of nonlinear stochastic synthetic genetic oscillators under intrinsic and extrinsic molecular noise via quorum sensing. BMC Syst Biol 6: 136.

Garcia-Ojalvo, J., Elowitz, M.B. and Strogatz, S.H. 2004. Modeling a synthetic multicellular clock: repressilators coupled by quorum sensing. Proc Natl Acad Sci USA 101: 10955–10960.

Takagi, T. and Sugeno, M. 1985. Fuzzy Identification of Systems and Its Applications to Modeling and Control. IEEE Transactions on Systems Man and Cybernetics 15: 116–132.

Wang, R. and Chen, L. 2005. Synchronizing genetic oscillators by signaling molecules. J Biol Rhythms 20: 257–269.

Wang, R., Chen, L. and Aihara, K. 2006. Synchronizing a multicellular system by external input: an artificial control strategy. Bioinformatics 22: 1775–1781.

Zhang, W., Chen, B.S. and Tseng, C.S. 2005. Robust H∞ filtering for nonlinear stochastic systems. IEEE Transactions on Signal Processing 53: 589–598.

Index

A

activator-regulated promoter-RBS
component 79, 80, 82, 118, 121
activator-regulated promoter-RBS library
69, 79–82
attenuation level 15, 17, 18, 20, 24, 25,
29–31, 34, 97–101, 104, 105, 107, 110,
113, 165, 172

B

BioBrick 3, 53, 55, 68, 75, 76, 79, 82, 111, 129
biological filter 5, 117–121, 123, 125–135

C

constitutive promoter-RBS component 69,
72, 74–76, 79, 118–123, 129, 130, 132,
133, 140, 145
constitutive promoter-RBS library 72–74,
76, 77, 79, 80, 129
cost function 87–89, 92, 127–129, 131, 132,
134, 146, 150, 152
crossover 40–44, 50, 54, 55, 58, 88, 89, 92,
128

D

design procedure 4, 15, 18, 22, 26, 27, 29, 31,
44, 55, 66, 84, 88, 89, 93, 94, 99, 102, 104,
105, 112, 128, 129, 135, 143, 146, 149,
152, 153, 165, 168, 171–174
design specification 3, 4, 9, 13–20, 24, 26–28,
30, 31, 39, 44, 53, 57, 60, 66, 87–89, 91,
93, 105, 125–131, 133, 144, 146, 152, 153

E

evolutionary algorithm (EA) 4, 38–44, 56,
58, 60
external disturbance 4, 14–25, 29–31, 33, 34,
39, 43, 45, 47, 55–58, 60

extrinsic robustness 155, 164, 165, 168–170

F

filtering ability 15–17, 19, 20, 22, 30, 31, 58,
98, 162–165, 167, 169, 178
fitness 4, 38–44, 50, 51, 55, 58, 60
fitness function 4, 39–42, 58, 60
fitness maximization 38, 40, 43

G

genetic algorithm (GA) 4, 38–40, 42–44, 47,
50–52, 54, 55, 60, 87–90, 92, 93, 127–129,
131, 146, 150, 152
genetic operator 40, 41, 88
genetic transistor 137–154
global linearization 4, 22–24, 26, 27, 29, 35

H

H_2 optimal reference tracking 98, 101, 105,
112
H_∞ noise attenuation level 98, 101, 104, 105,
107, 110, 113
Hamilton-Jacobi inequality (HJI) 20–22, 25,
100–102, 112–115, 155, 156, 163–165,
167, 168, 170, 171, 175, 176
Hill function 10, 27, 64, 77

I

I/O response 118, 119, 123, 125–137, 142,
145, 146, 150–153
initialization 41, 42
intrinsic parameter fluctuation 4, 5, 13–16,
18, 19, 23, 30, 31, 38, 45, 48–51, 55, 58,
87, 88, 93, 95–98, 105, 107, 111, 112, 128,
131, 134, 150, 152, 155, 164, 165, 169,
170, 172, 177
intrinsic robustness 155, 164, 165, 168–170
Ito stochastic differential equation 16, 96,
160

Color Plate Section

Chapter 3

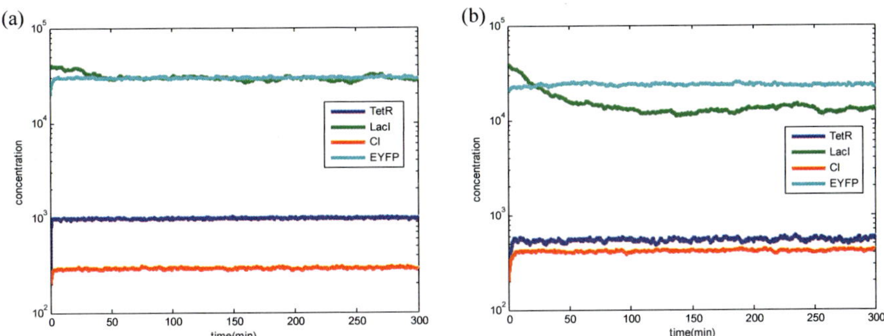

Figure 3.1. Simulation results for synthetic transcriptional cascade network. (a) With the design parameters in the specified parameter range given in. (b) With the design parameters outside the specified parameter range.

Chapter 4

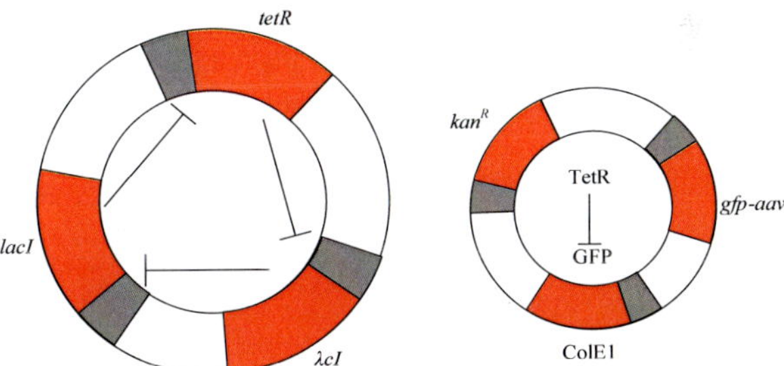

Figure 4.3. Construction of the repressilator network in the host cell, *E. coli*. The repressilator is a cyclic negative-feedback loop composed of three repressor genes (the red regions) *tetR*, *λcI*, *lacI* and their corresponding promoters (the gray regions) in plasmid. The compatible reporter plasmid expresses an intermediate-stability GFP variant (gfp-aav) on the ring.

(a)

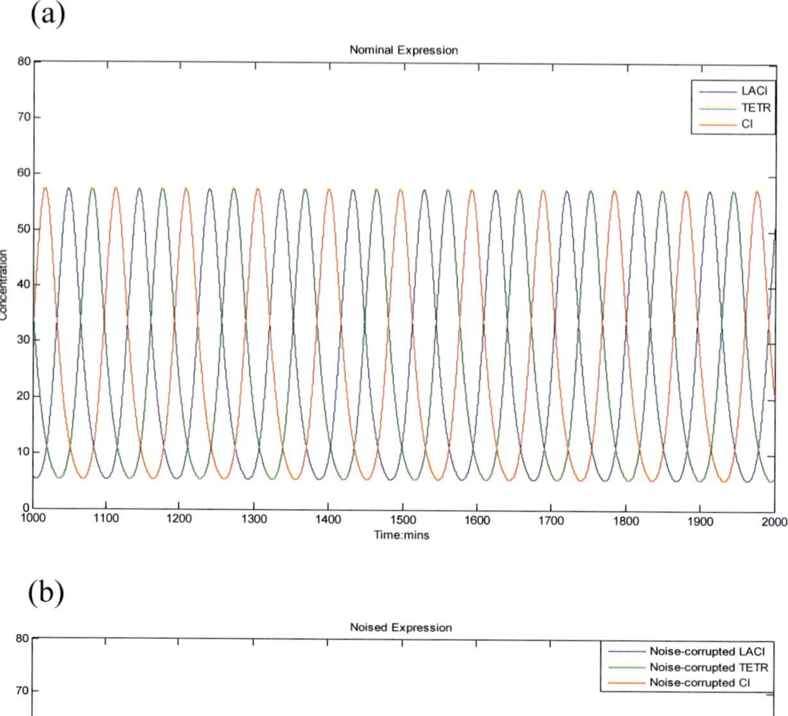

(b)

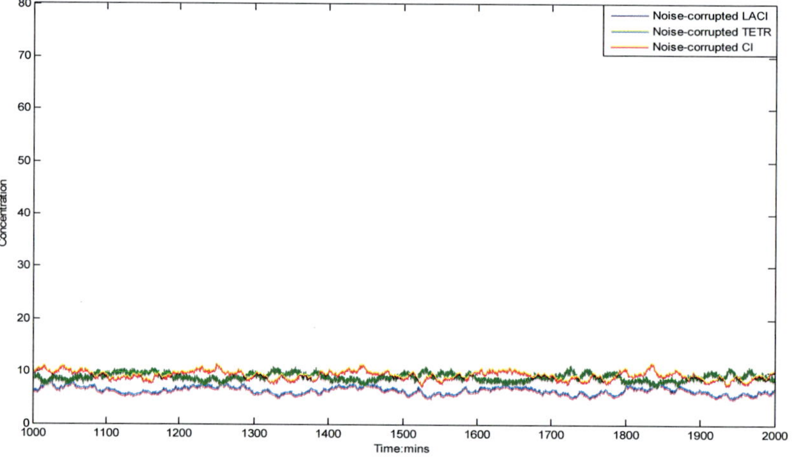

Figure 4.4. Time-responses of protein concentrations. (a) The nominal repressilator time-response with $\alpha_i = 0.5$, $\gamma_{pi} = 0.069$ for $i = 1,2,3$ by the minute in (Elowitz and Leibler 2000). (b) The repressilator time-response under intrinsic parameter fluctuations and extrinsic disturbances on the host cell. These two time-responses show that the repressilator in (Elowitz and Leibler 2000) suffers substantially from the effects of intrinsic parameter fluctuations and environmental noises on the host cell. Clearly, the corrupted repressilator does not have enough robustness to tolerate parameter fluctuations and extrinsic noises and loses its characteristics of oscillation.

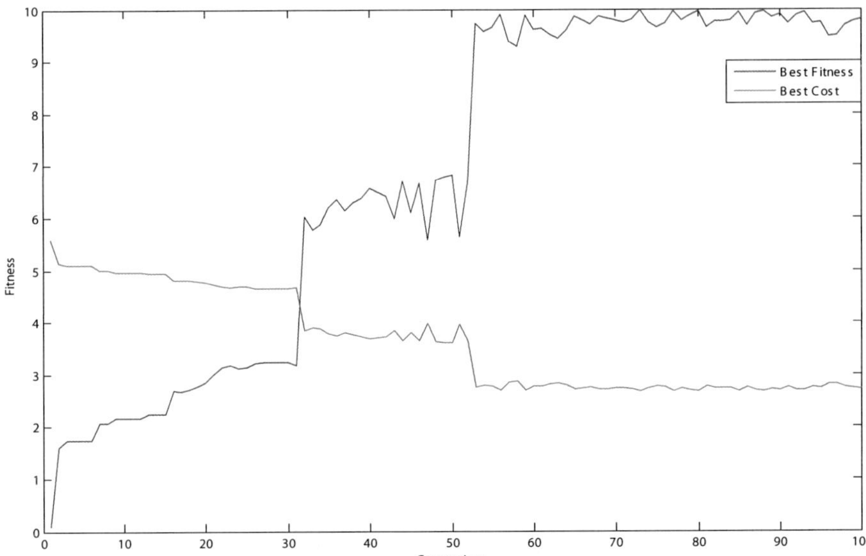

Figure 4.5. Convergence of fitness value. The best fitness score and best cost value evolve during the generations. The vibrations of the best cost value and the best fitness score come from the stochastic intrinsic parameter fluctuations and extrinsic noises, which fluctuate in each generation and directly affect the reliability of the synthetic gene network.

(a)

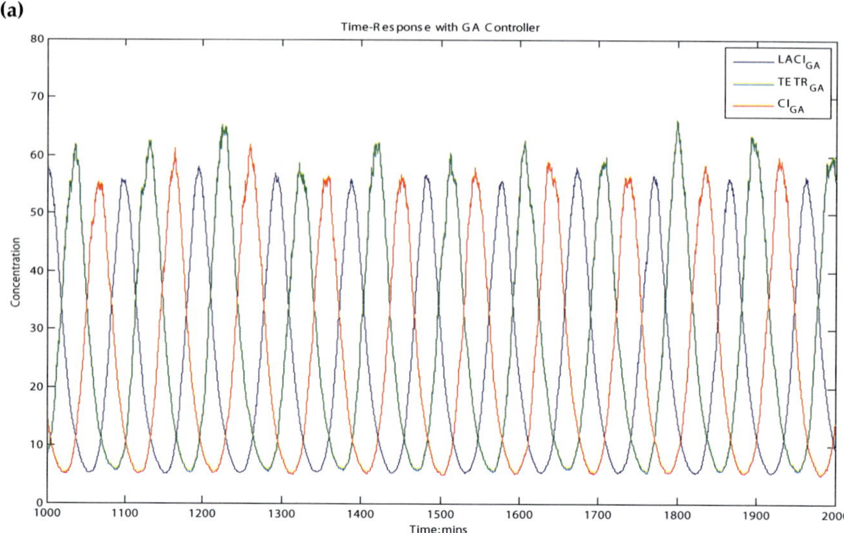

(b)

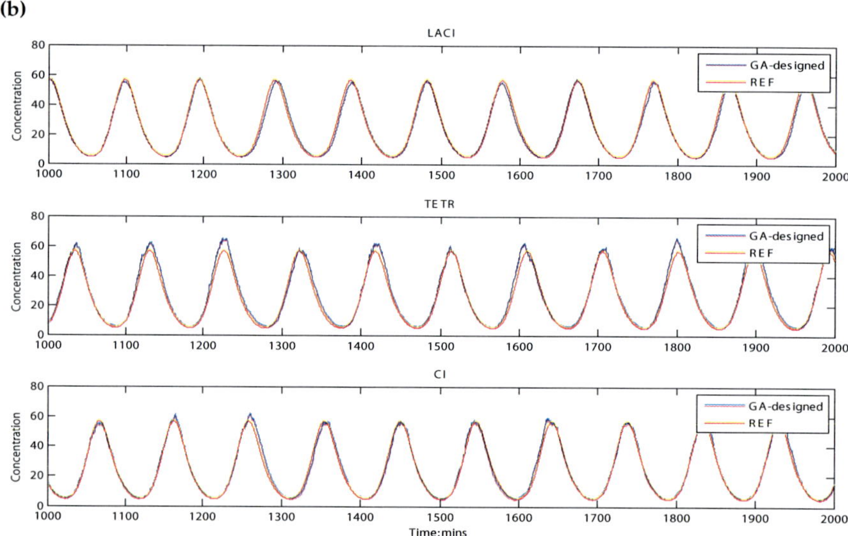

Figure 4.6. Time-response of the synthetic gene oscillator via the proposed GA-based design method solution. (a) Time-responses of these three proteins. (b) Time-response tracking of each protein and its reference. Under the parameter fluctuations and environmental noises, the designed repressilator can maintain its characteristics of oscillation and function properly. There are still some discrepancies between the desired reference signals and the protein concentrations of the repressilator, which are mainly due to parameter perturbations and environmental noises.

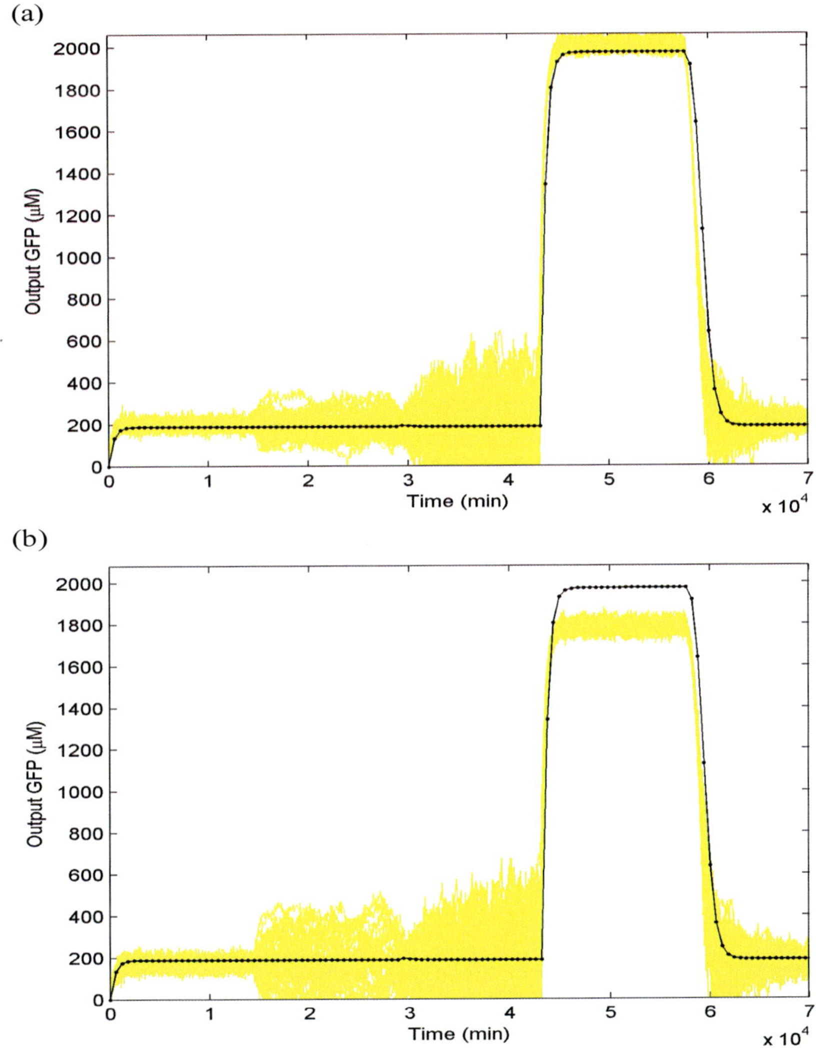

Figure 4.8. Simulation results of biological AND gate. (a) The simulation result with optimal design parameters $k_1^*=0.6042$, $k_2^*=0.8410$, $k_3^*=0.9272$, $k_4^*=0.2063$, $k_5^*=0.00001$, $k_6^*=0.8640$ and $k_7^*=0.1235$. The Monte Carlo simulations are performed by 50 rounds. The mean error is

$$\bar{e} = \frac{1}{50}\sum_{i=1}^{50} e_i = 121.95$$ with standard deviation of 9.93, where e_i is the root mean squared error of the ith Monte Carlo simulation. (b) In contrast to the above design case, the design parameters are specified aside the optimal design parameter k^*, with $k_1=0.9$, $k_2=0.6$, $k_3=0.95$, $k_4=0.25$, $k_5=0.00002$, $k_6=0.6$ and $k_7=0.15$. In this design case, the mean error is $\bar{e} = 165.90$ with standard deviation of 18.67.

Chapter 6

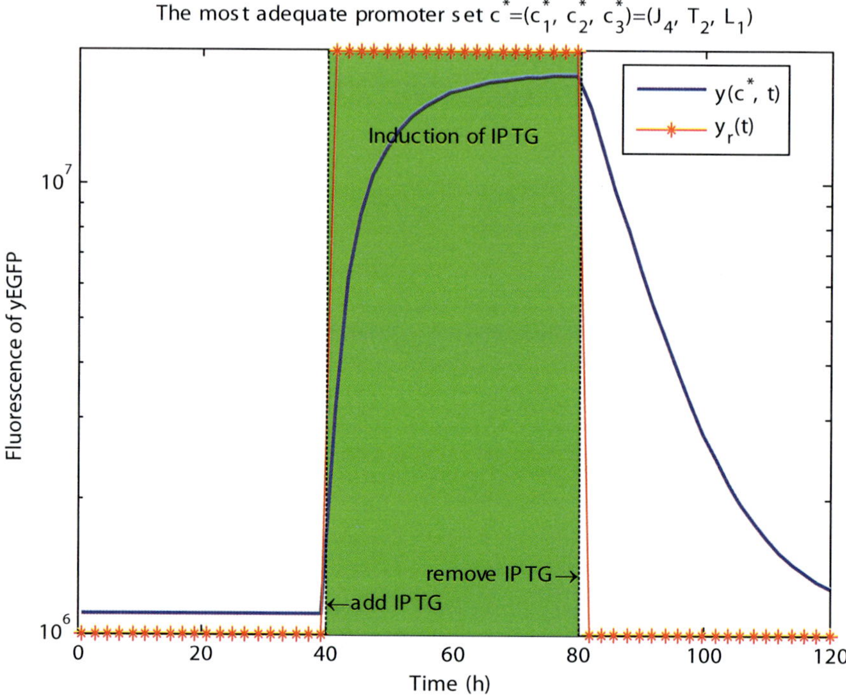

Figure 6.2. Simulation results for synthetic transcription cascade design. The most adequate promoter set $c^* = \{c_1^*, c_2^*, c_3^*\} = \{J_4, T_2, L_1\}$ is obtained via library-based search method through GA. The system behavior $y(c^*, t)$ of the gene network employs the most adequate promoter set c^* to track the desired reference trajectory $y_r(t)$, which is at high from 40 to 80 hours and at low in other hours. IPTG is added from 40 hours and removed from 80 hours.

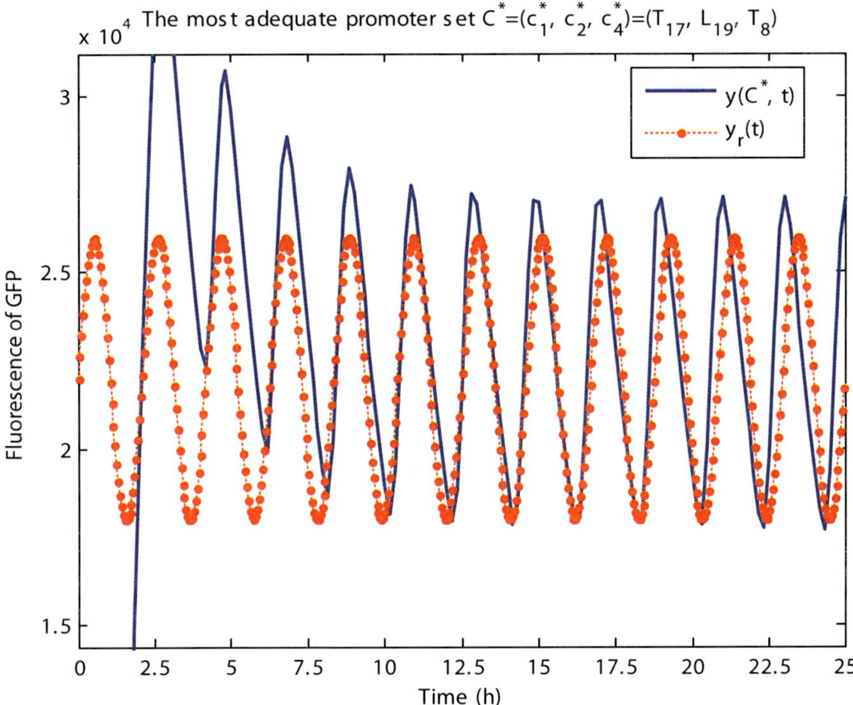

Figure 6.4. Simulation results for synthetic genetic oscillator design. The most adequate promoter set $c^* = \{c_1^*, c_2^*, c_4^*\} = \{T_{17}, L_{19}, T_8\}$ for the synthetic oscillator is selected using the library-based search method. The oscillatory behavior $y(c^*, t)$ with the most adequate promoter set c^* can track the desired reference trajectory $y_r(t)$.

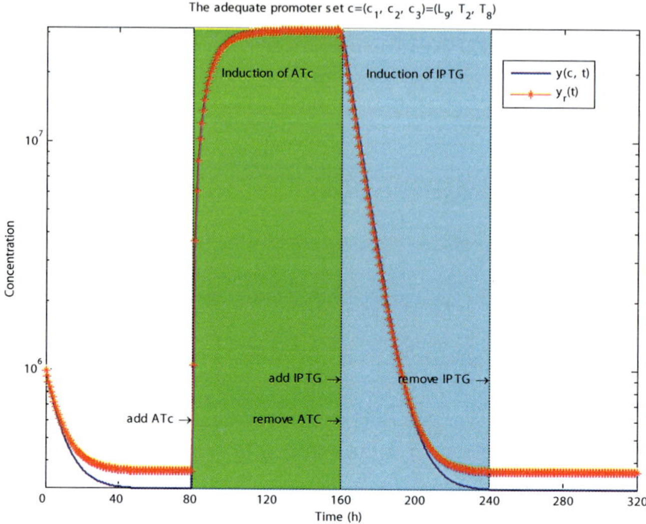

Figure 6.7. Simulation results for synthetic toggle switch design. By solving the LMIs-constrained optimization problem in (6.32) for the synthetic gene network in Figure 6.5, an adequate promoter set $c = \{c_1, c_2, c_3\} = \{L_9, T_2, T_8\}$ is selected from the corresponding promoter libraries. The synthetic gene network is added with inducer ATc to induce the gene network from 80 hours to 160 hours, and then is added with inducer IPTG from 160 hours to 240 hours. Obviously, the output $Y(c,t)$ can robustly track the desired reference output $Y_r(t)$.

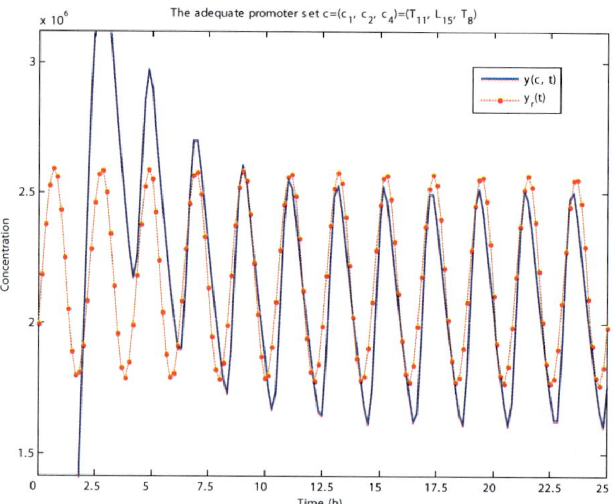

Figure 6.8. Simulation results for synthetic genetic oscillator design. The adequate promoter set $c = \{c_1, c_2, c_4\} = \{T_{11}, L_{15}, T_8\}$ is obtained via solving the multiobjective H_2/H_∞ reference tracking design problem in (6.32) for the synthetic gene network in Figure 6.3. The desired output $Y(c,t)$ of the gene network employs the adequate promoter set c to robustly and optimally track the desired behavior $Y_r(t)$ generated by the reference model. Our proposed design method could provide a genetic oscillator with the prescribed amplitude and period via selecting an adequate promoter set from the existing promoter libraries.

Chapter 7

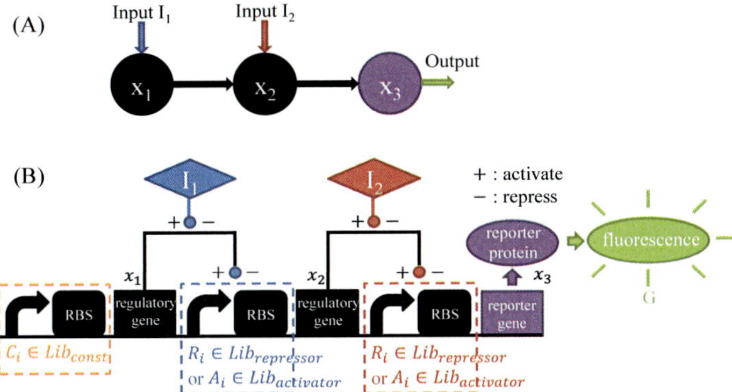

Figure 7.1. (A) A synthetic biological filter design based on a cascade gene circuit topology. (B) A synthetic cascade circuit with three promoter-RBS components is designed to function as a biological filter. The first node has been fixed by a constitutive promoter-RBS component. The second and third nodes can be engineered by an activator-regulated promoter-RBS component or a repressor-regulated promoter-RBS component, respectively. When different kinds of promoter-RBS components and regulatory genes are selected, they activate (+ sign) or repress (− sign) the connected node.

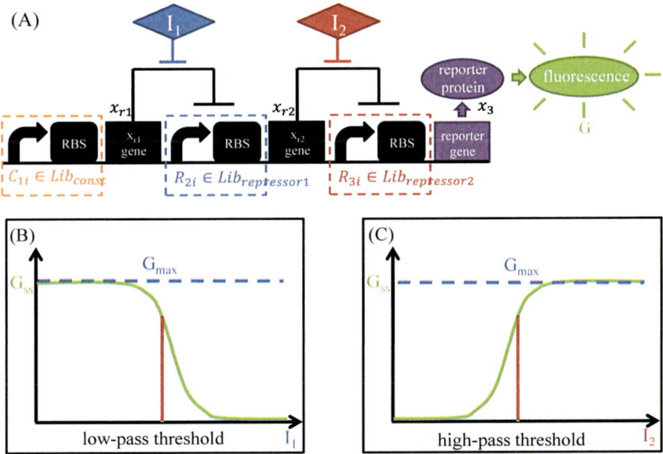

Figure 7.2. A synthetic biological filter with promoter-RBS components selected from the constitutive and repressor-regulated promoter-RBS libraries, and the I/O response of the biological low-pass and high-pass filters. (A) For convenience of explaining the operating mechanism of the biological filter, a synthetic biological filter is assembled by selecting a constitutive promoter-RBS component C_{1i}, and two different repressor-regulated promoter-RBS components R_{2i} and R_{3i} from their corresponding libraries, to regulate their corresponding regulatory genes and reporter gene. (B) The I/O response of the low-pass filter for input inducer I_1 and output fluorescence G_{ss} at steady state. (C) The I/O response of the high-pass filter for input inducer I_2 and output fluorescence G_{ss} at steady state.

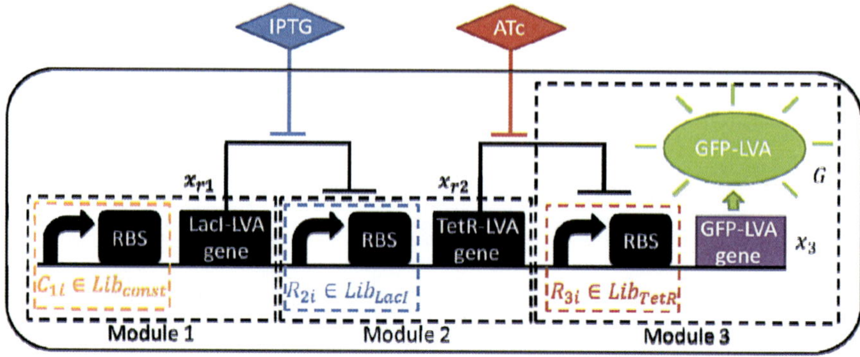

Figure 7.3. A design example of the biological low-pass filter. The biological low-pass filter can be divided into three cascade modules. Each module has both a promoter-RBS component selected from the corresponding promoter-RBS libraries and a specified gene downstream of the promoter. The first module contains a constitutive promoter-RBS component C_{1i} and a repressor LacI-LVA gene. The second module contains a LacI-regulated promoter-RBS component R_{2i} and a repressor TetR-LVA gene, and the third module contains a TetR-regulated promoter-RBS component R_{3i} and a reporter protein GFP-LVA gene.

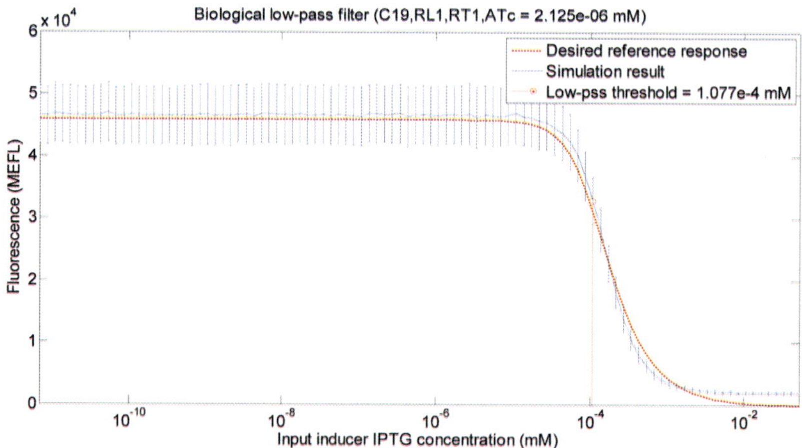

Figure 7.4. Monte Carlo simulation result of a biological low-pass filter displayed as the average of 1000 runs. By minimizing the cost function in (7.10) for the biological low-pass filter in Figure 7.3, the most adequate set $f_1^* = \{pR_1^*, pR_2^*, pR_3^*, I_2^*\} = \{C_{19}, R_{L1}, R_{T1}, ATc = 2.125 \times 10^{-6}\ \text{mM}\}$ is selected from the corresponding libraries. The blue solid line is the Monte Carlo simulation result displayed as the average of 1000 runs. The red dashed line is the desired I/O response generated by equation (7.8). The red solid line indicates the low-pass threshold of the biological filter.

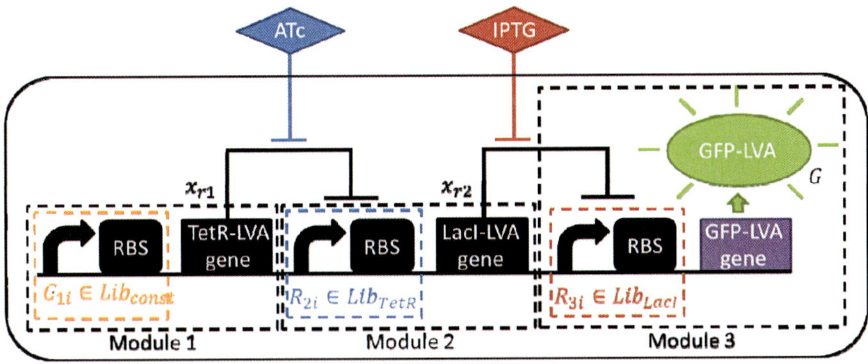

Figure 7.5. A design example of the biological high-pass filter. The biological high-pass filter can be divided into three cascade modules. The first module contains both a constitutive promoter-RBS component C_{1i} and a repressor TetR-LVA gene. The second module contains a TetR-regulated promoter-RBS component R_{2i} and a repressor LacI-LVA gene, and the third module contains a LacI-regulated promoter-RBS component R_{3i} and a reporter protein GFP-LVA gene.

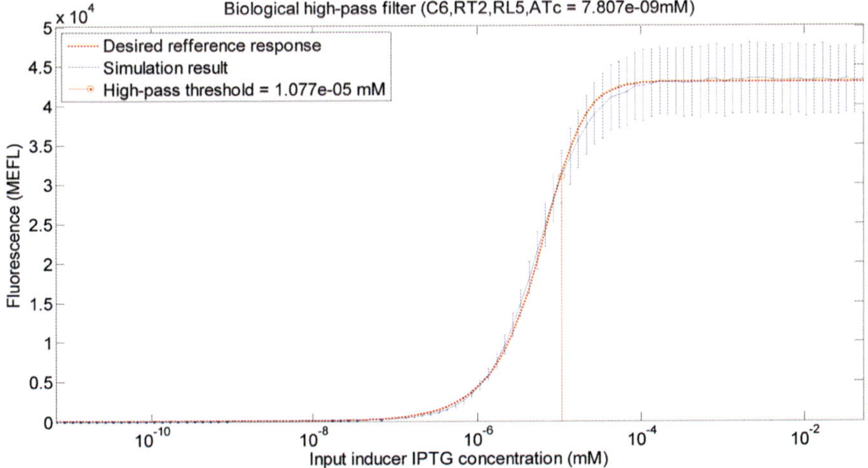

Figure 7.6. Monte Carlo simulation result of a biological high-pass filter displayed as the average of 1000 runs. By minimizing the cost function in (7.13) for the biological high-pass filter in Figure 7.5, the most adequate set $f_2^* = \left\{ pR_1^*, pR_2^*, pR_3^*, I_1^* \right\} = \left\{ C_6, R_{T2}, R_{L5}, ATc = 7.807 \times 10^{-9} \text{ mM} \right\}$ is selected from the corresponding libraries. The blue solid line is the Monte Carlo simulation displayed as the average of 1000 runs. The red dashed line is the desired I/O response generated by equation (7.12). The red solid line indicates the high-pass threshold of the biological filter.

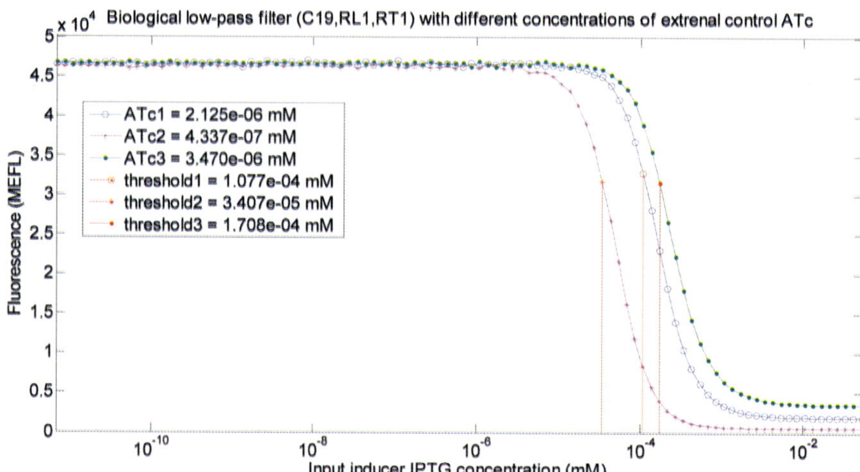

Figure 7.7. Monte Carlo simulations of external adjustability of a biological low-pass filter. The blue line is a low-pass I/O response with a specific threshold $u_{lc1} = 1.077 \times 10^{-4}$ mM and a maximum fluorescence expression $G_{max} = 4.702 \times 10^4$ MEFL based on the most adequate set $f_1^* = \{pR_1^*, pR_2^*, pR_3^*, I_2^*\} = \{C_{19}, R_{L1}, R_{T1}, ATc = 2.125 \times 10^{-6}$ mM$\}$ for the biological low-pass filter. The purple line is a low-pass I/O response with a smaller threshold $u_{lc2} = 3.407 \times 10^{-5}$ mM and a maximum fluorescence expression $G_{max} = 4.660 \times 10^4$ MEFL due to a decrease in the concentration of the external control ATc to 4.337×10^{-7} mM. The green line is a low-pass I/O response with a larger threshold $u_{lc3} = 1.708 \times 10^{-4}$ mM and a maximum fluorescence expression $G_{max} = 4.700 \times 10^4$ MEFL due to an increase in the concentration of the external control ATc to 3.470×10^{-6} mM.

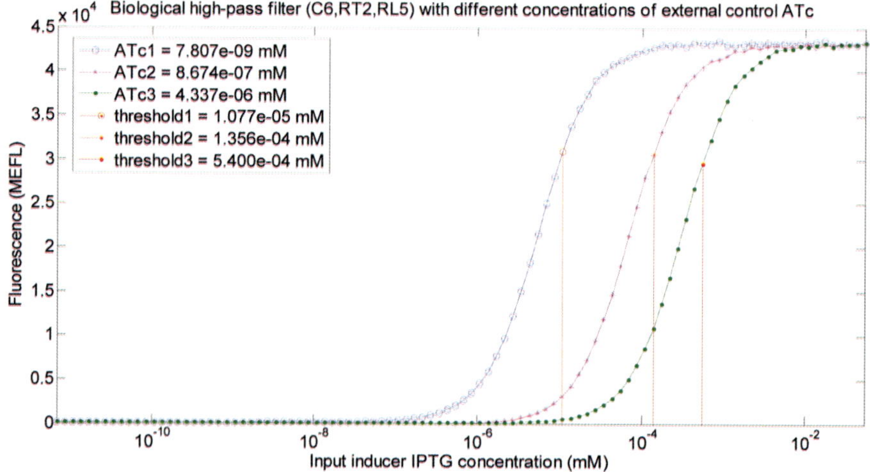

Figure 7.8 contd....

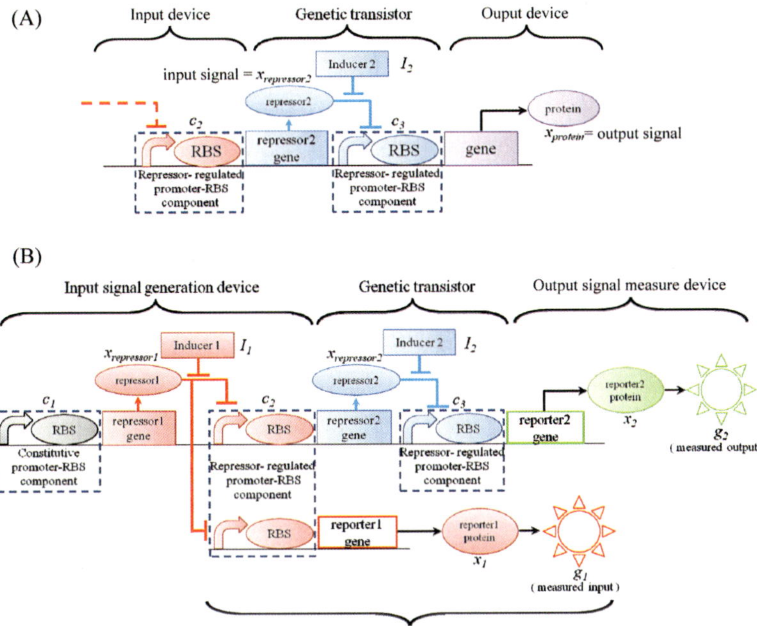

Figure 7.11. The representation of synthetic genetic transistor circuit with different measurement devices. (A) The synthetic genetic transistor consists of a repressor regulated promoter-RBS component c_3 and repressor2 gene $x_{repressor2}$ is the input signal and $x_{protein}$ is the output signal in the genetic transistor. (B) For measuring the I/O characteristics of synthetic genetic transistor circuit, the reporter1 gene and reporter2 gene are embedded at the downstream of repressor-regulated promoter-RBS component c_2 and the additional promoter-RBS component c_3 respectively. Then input protein concentration x_1 can be measured by input fluoresce by g_1 and output fluorescence g_2 of output protein concentration x_2 can be measured by output fluoresce g_2, respectively. In addition, for the convenience of controlling the input, the repressor1 gene is constructed with the constitutive promoter-RBS component c_1 to control the kinetic strength of c_2 and the corresponding inducer I_1 is used to control the fluorescence g_1 of input protein concentration x_1. It is found at steady state that input/output = $x_{repressor2}/x_{protein} \approx x_1/x_2 \approx g_1/g_2$. Therefore the I/O characteristic of $x_{repressor2}/x_{protein}$ in genetic transistor at steady state can be represented by the g_1/g_2 ratio.

Figure 7.8 (Facing page). Monte Carlo simulations of external adjustability of a biological high-pass filter. The blue line is a high-pass I/O response with a specific threshold $u_{hc1} = 1.077 \times 10^{-5}$ mM and a maximum fluorescence expression $G_{max} = 4.354 \times 10^4$ MEFL based on the most adequate set $f_2^* = \{pR_1^*, pR_2^*, pR_3^*, I_1^*\} = \{C_6, R_{T2}, R_{L5}, ATc = 7.807 \times 10^{-9} \text{ mM}\}$ for the biological high-pass filter. The purple line is a high-pass I/O response with a larger threshold $u_{hc2} = 1.356 \times 10^{-4}$ mM and a maximum fluorescence expression $G_{max} = 4.336 \times 10^4$ MEFL due to an increase in the concentration of the external control ATc to 8.674×10^{-7} mM. The green line is a high-pass I/O response with a much larger threshold $u_{hc3} = 5.400 \times 10^{-4}$ mM and a maximum fluorescence expression $G_{max} = 4.332 \times 10^4$ MEFL obtained by further increasing the concentration of external control ATc to 4.337×10^{-6} mM.

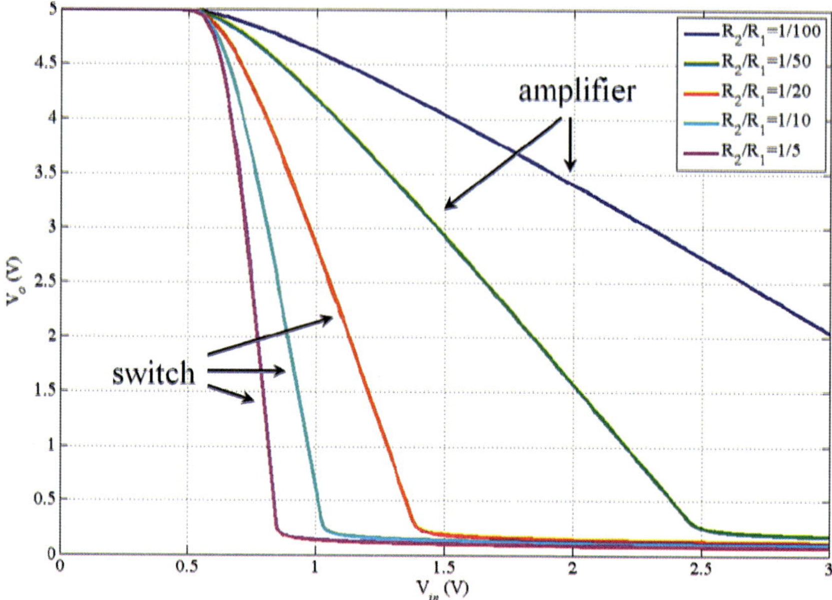

Figure 7.12. The voltage I/O characteristics of the common-emitter circuit for different R_2/R_1 ratio. The circuit is shown in Figure 7.9B, and simulated by the PSpice with a standard 2N3904 transistor from PSpice library. The voltage I/O characteristics are simulated by changing the R_2/R_1 ratio. According to (7.15), when the R_2/R_1 ratio becomes large, the voltage I/O characteristic is much sharper as the amplifier in linear region. And when the R_2/R_1 ratio is large enough, the voltage I/O characteristic will be like a switch.

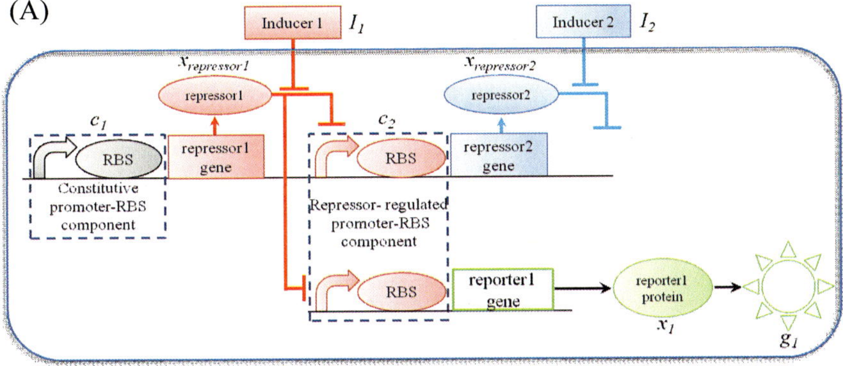

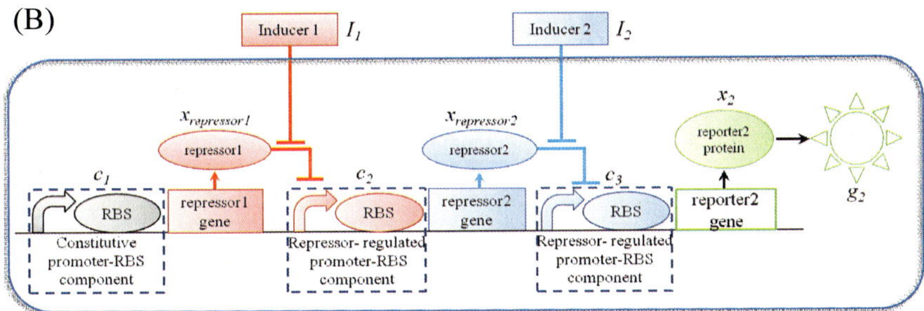

Figure 7.13. Input and output signal measurement of synthetic genetic transistor *in vivo*. For measuring the input fluorescence g_1 and output fluorescence g_2 with the same reporter gene, the synthetic genetic transistor circuit is divided into two parts in (A) and (B). These two parts will be constructed into different cells, respectively. Then, the cells will growth in the same condition and are measured at the same time. (A) The measurement of the input fluorescence g_1 of synthetic genetic transistor. (B) The measurement of the output fluorescence g_2 of synthetic genetic transistor.

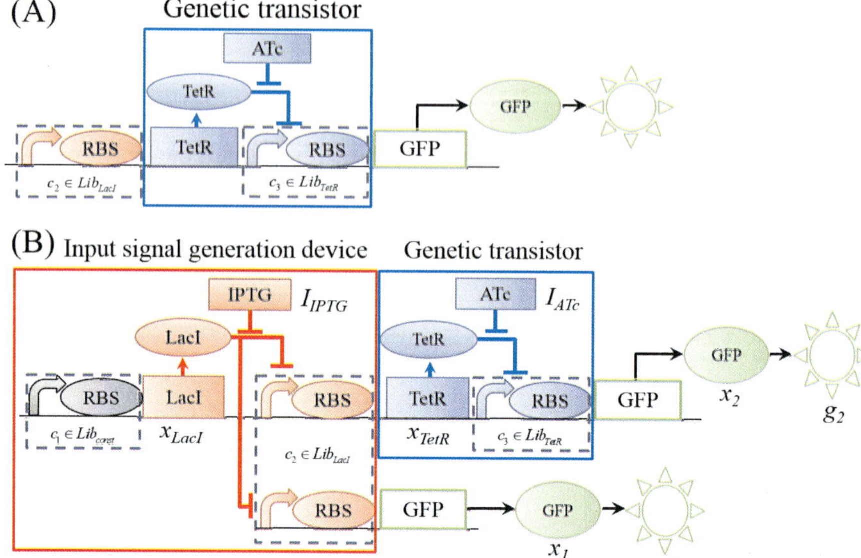

Figure 7.14. The genetic transistor design example based on promoter-RBS libraries. (A) The genetic transistor consists of repressor-regulated promoter-RBS component, $c_3 \in Lib_{Tet}$, TetR protein coding gene and inducer, ATc. (B) The complete genetic transistor with measurement circuit consists of three promoter-RBS components, c_1, c_2 and c_3, selected from promoter-RBS libraries, Lib_{const}, Lib_{LacI} and Lib_{TetR}, respectively. LacI protein x_{LacI} represses the strength of promoter-RBS component c_2 to decrease the concentration of TetR protein x_{TetR} and the fluorescence g_1 of GFP x_1. The inducer I_{IPTG} is added to bind LacI protein x_{LacI} to restrain LacI protein x_{LacI} from repressing the strength of promoter-RBS component c_2. And TetR protein x_{TetR} represses the strength of promoter-RBS component c_3 to decrease the fluorescence g_2 of GFP x_2. The inducer I_{ATc} is added to bind TetR protein x_{TetR} to restrain TetR protein x_{TetR} from repressing the strength of promoter-RBS component c_3.

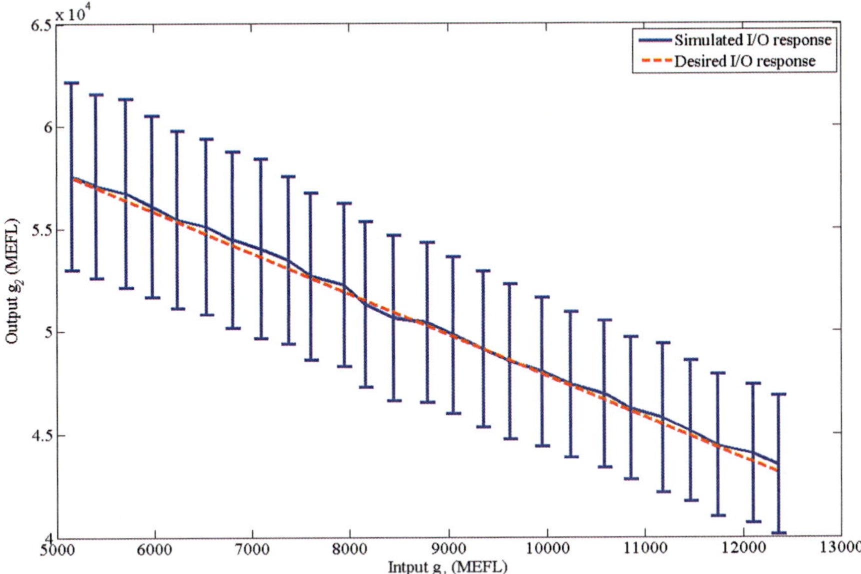

Figure 7.15. Monte Carlo simulation results of amplifier I/O response design example of synthetic genetic transistor. For amplifier design example of synthetic genetic transistor in Figure 7.14, the most adequate promoter-RBS component c_3 and ATc concentration I_{ATc} are searched as $\{c_3, I_{ATc}\} = \{T_3,\ 1.058\ \text{ng/ml}\}$ from the corresponding promoter-RBS library Lib_{TetR} and concentration range of inducer I_{ATc}. The Monte Carlo simulations are with 1000 runs. The blue line is the simulation result, and the error bars are the standard deviations. The red dash line is the desired I/O response generated by (7.29).

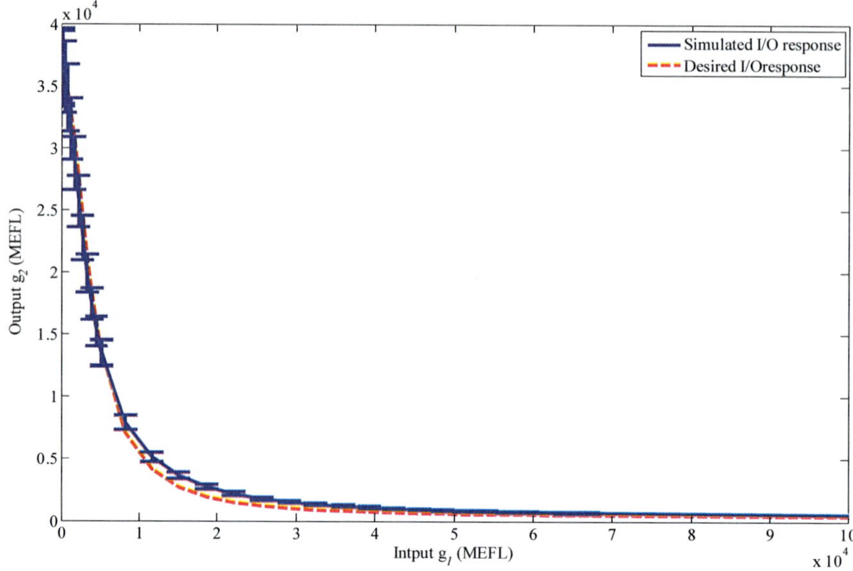

Figure 7.16. Monte Carlo simulations of switch design example of synthetic genetic transistor. For switch design example of synthetic genetic transistor in Figure 7.14, the most adequate promoter-RBS component c_3 and ATc concentration I_{ATc} are searched as $\{c_3, I_{ATc}\} = \{T_1,\ 0.182\ \text{ng/ml}\}$ from the corresponding promoter-RBS library Lib_{TetR} and concentration range of inducer I_{ATc}. The Monte Carlo simulations are with 1000 runs. The blue line is the simulation result, and the error bars are the standard deviations. The red dash line is the desired I/O response generated by (7.33).

Chapter 8

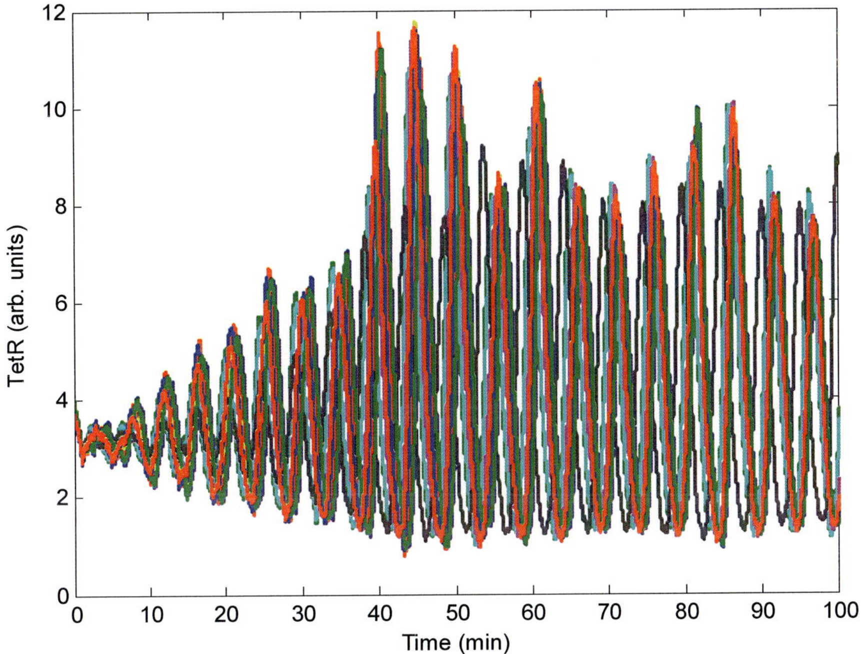

Figure 8.2. Ten coupled genetic oscillators. The parameter values in (8.1), (8.2), and (8.3) are set as follows: $\alpha_a=\alpha_b=\alpha_c=216$, $\alpha_S=20$, $\mu=1.2$, $\mu_S=1$, $n=2$, $\gamma_S=1$, $\eta_S=2$, $\beta_S=0.1$, $\beta_A=\beta_B=\beta_C=1$, $\gamma_m=6.9315$, $\gamma_p=1.1552$ and $Q_e=0.09$. Suppose the nonlinear stochastic coupled synthetic oscillators suffer from stochastic parameter fluctuations as shown in (8.5) with $\Delta\alpha_a=\Delta\alpha_b=\Delta\alpha_c=2.16$, $\Delta\alpha_S=0.2$, $\Delta\beta_A=\Delta\beta_B=\Delta\beta_C=0.01$, $\Delta\beta_S=0.001$, $\Delta\eta_S=0.02$, $\Delta\gamma_m=0.06$, $\Delta\gamma_p=0.01$, and $\Delta\gamma_S=0.01$. For the convenience of simulation, we assume that the extrinsic molecular noise $v_1 \sim v_{10}$ is independent Gaussian white noise with a mean of zero and standard deviation of 0.02. It can be seen that coupled synthetic oscillators cannot achieve synchronization under these intrinsic kinetic parameter fluctuations and extrinsic molecular noise.

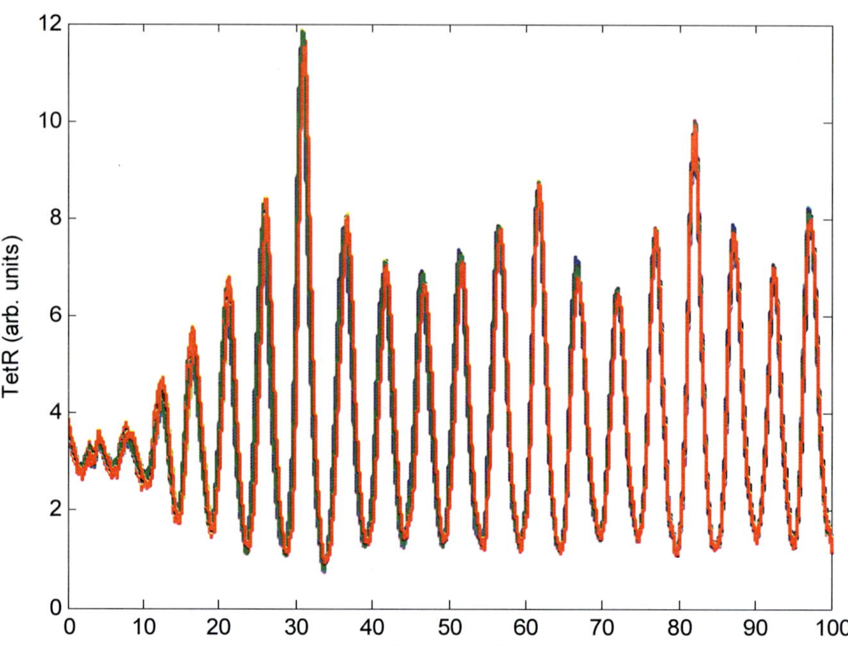

Figure 8.3. Robust synchronization result of ten coupled synthetic oscillators in Figure 8.2, by external control with $Q = 0.66$.